THE QUANTUM MECHANICS OF
MINDS AND WORLDS

√B

6/00

The Quantum Mechanics of Minds and Worlds

JEFFREY ALAN BARRETT

OXFORD
UNIVERSITY PRESS

OXFORD
UNIVERSITY PRESS

Great Clarendon Street, Oxford OX2 6DP

Oxford University Press is a department of the University of Oxford.
It furthers the University's objective of excellence in research, scholarship,
and education by publishing worldwide in

Oxford New York

Athens Auckland Bangkok Bogotá Buenos Aires Calcutta
Cape Town Chennai Dar es Salaam Delhi Florence Hong Kong Istanbul
Karachi Kuala Lumpur Madrid Melbourne Mexico City Mumbai
Nairobi Paris São Paulo Singapore Taipei Tokyo Toronto Warsaw

with associated companies in Berlin Ibadan

Oxford is a registered trade mark of Oxford University Press
in the UK and certain other countries

Published in the United States
by Oxford University Press, Inc., New York

British Library Cataloging in Publication Data
Data available

Library of Congress Cataloging in Publication Data

Barrett, Jeffrey Alan.
The quantum mechanics of minds and worlds / Jeffrey Alan Barrett
Includes bibliographical references and index.
1. Quantum theory. 2. Physical measurements. I. Title.
QC174.12.B364 1999 530.12–dc21 99–28680

ISBN 0-19-823838-X

1 3 5 7 9 10 8 6 4 2

Typeset by
Newgen Imaging Systems (P) Ltd., Chennai, India
Printed in Great Britain
on acid-free paper by
Biddles Ltd, Guildford and King's Lynn

For Martha, Thomas, and Jacob

Alexander wept when he heard from Anaxarchus that there was an infinite number of worlds; and his friends asking him if any accident had befallen him, he returns this answer: 'Do you not think it a matter of lamentation that when there is such a vast multitude of them, we have not yet conquered one?'

(Plutarch, *On the Tranquillity of Mind*)

PREFACE

T H I S book is about the quantum measurement problem, Hugh Everett III's proposed resolution, and some of the attempts to understand how it was supposed to work. While there is a brief review of the standard formulation of quantum mechanics (complete with a description of a two-slit experiment!) and a short appendix describing the Hilbert-space formalism, it is assumed that the reader already knows something about how quantum mechanics works and is comfortable with at least some of the mathematics. There is, in my opinion, no better introduction to the ways of quantum mechanics than David Albert's book *Quantum Mechanics and Experience*. One might also want to work through a careful presentation of the theory that includes a more detailed description of the mathematical formalism. P. A. M. Dirac's *Principles of Quantum Mechanics* is the classic introductory text (and I use Dirac's notation throughout this book). My favourite advanced introduction is Gordon Baym's *Lectures on Quantum Mechanics*. John Wheeler and W. H. Zurek's *Quantum Theory and Measurement* is the standard anthology on the measurement problem. I have tried to refer to page numbers in this anthology whenever possible.

Many conversations with friends and colleagues contributed to this book; in particular, I should like to thank Wayne Aitken, Frank Arntzenius, Guido Bacciagalluppi, Jeffrey Bub, Rob Clifton, Michael Dickson, Richard Healey, Meir Hemmo, Peter Lewis, Barry Loewer, Pen Maddy, Brad Monton, Laura Reutsche, Simon Saunders, and Brian Skyrms. I am especially indebted to David Albert for many enlightening discussions over the past several years—anyone familiar with Albert's work will immediately recognize his influence on the way that I think about quantum mechanics. I should also like to thank the anonymous referees who read this book in manuscript form—their comments were invaluable in putting together the final version. Finally, I should like to thank Ryan Barrett, whose excellent work produced the final figures. It was a pleasure working with the Oxford University Press editors—they were careful, smart, and patient.

This book was supported by a University of California President's Research Fellowship, and most of it was written while I was a Visiting Fellow at the University of Pittsburgh Center for the History and Philosophy of Science in 1996–7. I should like to thank both universities for their kind support.

J. A. B.

CONTENTS

LIST OF FIGURES

A BRIEF INTRODUCTION

T H E standard theory of quantum mechanics, as formulated by P. A. M. Dirac and John von Neumann, is in one sense the most successful physical theory ever—no other theory has ever made such accurate empirical predictions. It is all the more impressive because what it successfully predicts, the behaviour of the basic constituents of all physical things (electrons, protons, neutrons, photons, etc.), is often wildly counter-intuitive. There is, however, a problem. If one tries to understand the standard formulation of quantum mechanics as providing a complete and accurate framework for the description of all physical interactions, then it soon becomes evident that the theory is at least ambiguous, and, on a less charitable reading, one might even conclude that it is logically inconsistent. This is known as the quantum measurement problem. Hugh Everett III's formulation of quantum mechanics and the various reconstructions of his theory that have appeared since are all attempts to solve the measurement problem. But before considering possible solutions to the measurement problem, it is important to be clear about exactly what the problem is.

The basic constituents of matter, when left to themselves, behave in a way that is apparently nothing like the behaviour of the middle-sized objects (chairs, coins, cars, cats, etc.) that form the bulk of our experience and the basis for our physical intuitions. Because their behaviour is so counter-intuitive, any empirically adequate theory, any theory that makes the right empirical predictions for the experiments that we have performed so far, is bound to be itself counter-intuitive. The standard theory of quantum mechanics is certainly counter-intuitive. But that is not the problem. Rather, the problem is that the standard theory cannot be taken to provide a complete and accurate physical description of the odd behaviour that it is supposed to describe.

Our most careful observations suggest that the basic constituents of matter behave in a fundamentally random way. They also suggest that the basic constituents of matter sometimes behave like particles and sometimes like waves. The particle-like behaviour of fundamental particles is seen in such phenomena as cloud-chamber tracks and marks on photographic film. This particle-like behaviour agrees well with the physical intuitions we have developed from our experience with middle-sized physical systems. The wave-like behaviour of fundamental particles (and of other simple, well-isolated physical systems) is seen in interference phenomena. This wave-like behaviour of matter is very different from what one would expect from middle-sized physical systems—one might, for example, expect a particle (or any other physical object) to have a determinate position and to follow a determinate trajectory, not to spread out like a wave on a pond.

While one might lament the loss of classical determinism, it is the dual behaviour of matter that is really puzzling. Particles (and other simple, well-isolated systems) seem to behave one way when no one is looking (the odd quantum wave-like way) and another way when someone is. This dual behaviour is represented in the standard formulation of quantum mechanics by two dynamical laws: one law describes the evolution of a physical system when no one is looking and the other describes the evolution of the system when someone does. These two dynamical laws and the criterion for when each obtains is the ultimate source of the measurement problem in the standard theory.

There are two stock examples of quantum interference effects that we will return to in various forms throughout the book. One of these is the two-slit experiment and the other is Wigner's Stern–Gerlach experiment. Both are discussed below, followed by a somewhat more exotic example. Each experiment shows the sort of quantum weirdness that any satisfactory formulation of quantum mechanics must ultimately predict and explain.

1.1   *A textbook example: The two-slit interference experiment*

In the standard two-slit experiment one imagines launching particles, one at a time, at a barrier with two slits cut in it and with a screen beyond the slits (Fig. 1.1). This sort of experiment works with fundamental particles like electrons, but similar experiments have been performed with larger, more complicated systems. Indeed, atoms, molecules, superconducting rings, and other physical systems exhibit similar interference effects in the context of other experiments. Most of the particles just crash into the barrier, but some pass through the slits and strike the screen. It is the pattern of the particle distribution on the screen that suggests that the particles behave like waves while they are travelling towards the screen and no one is looking.

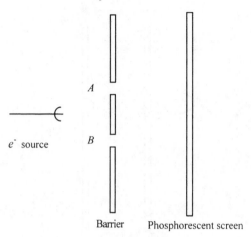

e^- source

FIG. 1.1 Two-slit setup.

First, consider what happens if we force each particle to pass one at a time through either one slit or the other. Suppose we close the lower slit *B* so that each particle must pass through the upper slit *A*. In this case, unsurprisingly, we get a particle distribution on the screen that is perfectly consistent with each particle having passed through *A* (one determines the particle distribution by choosing a convenient partition of the screen, then counting the number of particles that hit in each cell of the partition). Call this the *A* distribution. If we close *A* and send one particle at a time towards the barrier, we get a particle distribution that is perfectly compatible with each particle passing through *B*. Call this the *B* distribution.

So if we open both slits and send each particle one at a time towards the barrier, then if each particle passes through either *A* or *B*, one would expect to get a particle distribution that is a weighted sum of the *A* distribution (Fig. 1.2*a*) and the *B* distribution (Fig. 1.2*b*). If we aim the particles at a point halfway between the two slits (as well as we can aim them), then one would expect that about as many particles would pass through *A* as would pass through *B*, so one would expect to get a distribution like Figure 1.3*a*. Call this the *A*-or-*B* distribution—this is the distribution one would expect to get with a symmetric source if each particle either determinately passed through *A* or determinately passed through *B*.

We can check to make sure that the *A*-or-*B* distribution is what we do in fact get when each particle determinately goes through *A* or determinately

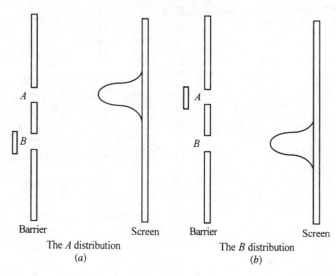

FIG. 1.2   *A* open and *B* open. (*a*) The *A* distribution. (*b*) The *B* distribution.

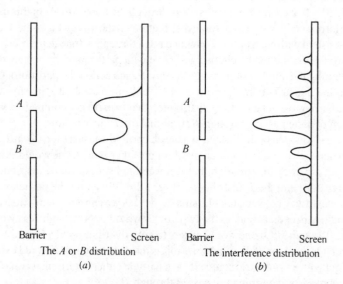

FIG. 1.3   What should happen and what does happen. (*a*) The *A*- or *B*- distribution. (*b*) The interference distribution.

goes through *B* by sending the particles one at a time toward the barrier and then closing one or the other of the two slits at random for each trial. When we do this, we get precisely the *A*-or-*B* distribution that one would expect (Fig. 1.3*a*). Consequently, one might naturally expect that if we do not move the source but open both *A* and *B* at the same time, then we will again get the *A*-or-*B* distribution. Indeed, one might be tempted to claim that with a symmetric source each particle determinately passes through one or the other of the two slits if and only if one gets the *A*-or-*B* distribution.

So what distribution do we get when both slits are open at the same time? We get one like Figure 1.3*b*. Call this the interference distribution, *and note that it is nothing like the A-or-B distribution*. The immediate consequence is that if one takes the *A*-or-*B* distribution to be characteristic of each particle either determinately passing through *A* or determinately passing through *B*, then one must conclude on the basis of what we actually see when we perform such experiments that at least some particles did not determinately pass through either slit! These particles did, however, somehow manage to get to the screen. So how did they get to the screen without determinately passing through *A* or determinately passing through *B*? What sort of trajectories could they have followed?

According to the standard von Neumann–Dirac formulation of quantum mechanics (which we will consider in some detail in the next chapter) the particles do not determinately pass through one slit or the other when both *A* and *B* are open; rather, the theory tells us that each particle follows a *superposition* of different trajectories and thus ends up in a superposition of passing through *A* and passing through *B*. A particle in a superposition of passing through each slit does not determinately pass through *A*, does not determinately pass through *B*, does not determinately pass through *A* and determinately pass through *B*, and does not determinately not pass through *A* and determinately not pass through *B*. Indeed, a particle that is in a superposition of passing through *A* and *B* has *observable* physical properties that differ from each of these four classical alternatives. It is the fact of each particle's following a superposition of trajectories rather than determinately passing through one slit or the other that explains the interference distribution.

While a particle is moving toward the screen, one might think of it as a wave, and just as it would be a mistake to ask for the single precise position of a wave, the standard theory tells us that the particle typically fails to have a determinate position. Part of the wave passes through *A* and part passes through *B*. These two wave packets spread out and interfere with each other in the region between the barrier and the screen, and then the composite wave strikes the screen. At approximately this point in the story, and one cannot say exactly when or how this occurs, the particle stops acting like a wave—when we look at the screen we do not see the global effects one would expect from a spread-out wave striking the surface of the screen but rather we see the effects one would expect from a single particle striking a single determinate point. On the other hand, the wave-like behaviour of each particle individually is seen in the overall particle distribution when the basic experiment is repeated many times. One might naturally conclude from such repeated experiments that the wave associated with each particle determines the *probability* of that particle being found in each region of the screen, and this is precisely what the standard theory predicts (and, to some extent, explains).

The state of a physical system is represented by a vector in quantum mechanics (something I shall discuss in more detail in the next chapter and Appendix A). If we represent the state where a particle P determinately passes through A by the vector $|A\rangle_P$ and the state where it determinately passes through B by the vector $|B\rangle_P$, then a state where P is in a superposition of passing through each slit might be represented by a linear combination (or sum) of these two vectors $\alpha|A\rangle_P + \beta|B\rangle_P$, where α and β are complex-valued coefficients. When one is interested in the position of a particle, the vectors one uses to represent the particle's state are functions that assign a complex number to each possible position. The wave function ψ that represents the state of a particle may be spread out or well-localized. If it is spread out, then the particle is far from having a determinate position. The better localized the wave function, the closer the particle is to having a fully determinate position.

The puzzle is that, regardless of a particle's initial state, whenever one looks for it (that is, whenever one makes a position measurement), one always finds it with a determinate position. The von Neumann–Dirac formulation of quantum mechanics explains this by stipulating that whenever one looks for a particle, that particle's state instantaneously and randomly *collapses* to a state where the particle has a determinate position (is in an eigenstate of position); that is, the particle's initially spread-out wave function instantaneously and randomly evolves to a wave function that is localized to the region where one found the particle (more generally, the particle's state randomly evolves to a state where whatever physical property one is observing has a determinate property—it evolves to an eigenstate of the observed property). Which determinate position the particle acquires is given by probabilities determined by the wave function that represented the particle's state just before it was observed. More specifically, the standard theory predicts that if ψ represents the complex-valued wave associated with a particular particle, then the probability of finding that particle in a region R is equal to the sum (or, more precisely, the integral) of $|\psi|^2$ over R.

Before we look at the screen, the standard theory tells us that there is no determinate matter of fact concerning where the particle is. According to the standard interpretation of states (given in the standard theory by the eigenvalue–eigenstate link),[1] it is not that the particle has a determinate position and that we do not know what it is; rather, it is that it simply fails to have any determinate position at all before it is observed. If each particle had a determinate but unknown position, then one would presumably get the A-or-B distribution whenever both slits were open, since each particle would in fact determinately pass through one slit or the other; but since we do not get the A-or-B distribution, this was taken as evidence that the particles do not pass through determinate slits, which is precisely what the standard theory tells us. The probabilities predicted by the standard theory, then, are not a matter of our ignorance—rather, being in a superposition of different states is observationally distinguishable from determinately being in one or the other of the states. In particular, a system in a superposition of states exhibits interference effects not exhibited by a system determinately in one of the elements of the superposition. But while each particle in the two-slit experiment fails to have a determinate position just before it strikes the screen, the standard theory tells us that the process of observing the particle somehow endows it with a determinate position. Further, the standard theory tells us that if one looks to see which slit each particle passes through, then one will cause the wave function associated with each particle to collapse to an eigenstate of passing through one or the other of the two slits (a state where the particle either determinately passes through A or determinately passes through B). And, sure enough, whenever we look to see which slit each particle passes through, the interference effects are destroyed and we get the A-or-B distribution. A proponent of the standard collapse theory would take this as (further) evidence for supposing that the usual wave dynamics is suspended and the state of a system randomly collapses to an eigenstate of the observable

---

[1] The eigenvalue–eigenstate link says that a system has a particular physical property if and only if its state is represented by an eigenvector of the operator representing the property.

one measures whenever a measurement is made (and I shall discuss all this in more detail in the next chapter).

It is interesting to consider what one would have to do in order to account for what we see if one insists, contrary to the standard theory, that each particle does in fact determinately pass through A or determinately pass through B in the two-slit experiment above. Since the behaviour of a particle passing through A depends on whether B is open at the instant the particle passes the barrier (since whether B is or is not open determines whether the particle passing through A will behave according to the A-or-B statistics or the interference statistics, and the other way round), the particle would have to have some way of learning whether B was open when it passed through A. That is, in order to say that each particle determinately passes through one or the other of the two slits, one would somehow have to explain how the state of the slit that it did *not* pass through, whether it was open or closed, affects the particle's motion even when there are no other particles around to carry the information about the state of that slit—remember: all this happens when we send the particles through the apparatus *one at a time*.

One strategy would be to postulate the existence of a new type of field that affects the motion of each particle in a way that depends on the state of the slit that the particle did *not* pass through. If there were such a field, then one could say that each particle determinately passed through one slit or the other and then account for the fact that a particle's behaviour typically depends on the state of the slit that they did not pass through by appealing to differences in the field. While all this is flatly inconsistent with the standard formulation of quantum mechanics, this is essentially how David Bohm's hidden-variable formulation of quantum mechanics works. I shall discuss Bohm's theory in more detail later. The point of mentioning it here is just to illustrate the sort of story one would have to tell in order to explain interference effects if one insisted, in accord with our intuitions but contrary to the standard theory, that each particle determinately passes through one slit or the other in the two-slit experiment above.

1.2 *Another textbook example: Spin properties of spin-$\frac{1}{2}$ systems*

Particles can be in superpositions of being located at different positions, but they can also be in superpositions of having other incompatible properties. A particle can, for example, be in a superposition of having different momenta, energies, or different spin properties. The spin properties of spin-$\frac{1}{2}$ particles provide one of the simplest examples of the nonclassical behaviour of matter.[2]

Electrons (and other spin-$\frac{1}{2}$ particles) exhibit a property called x-spin (spin-$\frac{1}{2}$ particles can be x-spin up or x-spin down) and another property called z-spin (they can be z-spin up or z-spin down). The results of x-spin and z-spin measurements are repeatable: if one measures the x-spin of an electron and finds it to be in an x-spin down state, then if one makes a second x-spin measurement (without disturbing the electron between measurements), one would again find it to be x-spin down.

There is a special relationship between x-spin and z-spin. If one finds that an electron is x-spin down, and then measures its z-spin, one gets each of z-spin up and z-spin down about half of the time. If one in fact gets z-spin down and then *remeasures* its x-spin, one gets each of x-spin up and x-spin down about half of the time. One might naturally con- clude that an intervening z-spin measurement disturbs the x-spin of an electron, but the odd thing here is that this disturbance does not come in degrees—if one starts with an electron in an x-spin up eigenstate (a state where the electron is determinately x-spin up) then measures the z-spin of the electron, the result of a subsequent x-spin appears to be completely random regardless of how careful one is in making the intervening z-spin measurement. Similarly, x-spin measurements appear to randomize com- pletely the z-spin of an electron. It seems then that there is a conspiracy that somehow prevents us from ever simultaneously knowing both the x-spin and the z-spin of an electron.[3] If spin properties behaved in the old-fashioned, classical, common-sense way of middle-sized objects, then one would expect no such epistemic limits; rather, one would expect that careful enough measurements would allow one to know both the x-spin and the z-spin of an electron.

Now consider an experiment where we send electrons one at a time into an x-spin sorting device. The sorting device works as follows: if an x-spin up electron enters the device, then it follows path A to I; and if an x-spin down electron enters the device, then it follows path B to I (Fig. 1.4). We can check to make sure the x-spin sorter is working by measuring the x-spin of particles on each path. When we look, we find that each particle

---

[2] See Baym (1969: 302–46) for a more complete discussion of spin-$\frac{1}{2}$ systems.

[3] This is an example of a quantum uncertainty relation: x-spin and z-spin are related to each other much as position and momentum are related in the Heisenberg uncertainty relation.

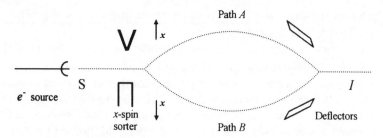

FIG. 1.4 Two-path setup.

takes exactly one of the two paths and that all the electrons on path *A* are *x*-spin up and that all the electrons on path *B* are *x*-spin down.

Suppose we send *z*-spin up electrons into the *x*-spin sorter: what should the statistics be for a *z*-spin measurement at *I*? One might expect about half of the electrons to take path *A* and about half to take path *B*. Electrons taking path *A* would be *x*-spin up and hence about half would be found to be *z*-spin up and half *z*-spin down. Similarly, electrons taking path *B* would be *x*-spin down and hence about half would be found to be *z*-spin up and half *z*-spin down. And we find precisely these statistics when we measure electrons on each of the two paths. So, if we assume that each electron that we send into the *x*-spin sorter either determinately takes path *A* or determinately takes path *B*, then about half of the electrons at *I* should be *z*-spin up and about half should be *z*-spin down. But this is not at all what happens. Whenever such an experiment is actually performed, *all* of the electrons measured at *I* are found to be *z*-spin up!

Again we have a result that seems to be incompatible with the assumption that each electron took one or the other of the two paths, but we know that each electron somehow gets to *I*; and whenever we look, we always see that each electron is on one path or the other, *and our looking at the paths causes the statistics to change to half up and half down at I*. The situation is made even more puzzling by the fact that the statistics become half *x*-spin up and half *x*-spin down at *I* if we watch either path, no matter how careful we are and *even if we find that no electron in fact travelled the path we were watching*. Similarly, the presence of a barrier on either path affects the statistics at *I* (it changes them from all *z*-spin up to half up and half down) even if we look and find that no electron has interacted with the barrier. It seems wrong, then, to say that each electron arriving at *I* takes one or the other of the two paths, but it also seems wrong to say that each electron follows both paths or no path at all.

According to the standard theory, because a z-spin up particle P is in a superposition of x-spin up and x-spin down eigenstates ($|\uparrow_z\rangle_P = 1/\sqrt{2}(|\uparrow_x\rangle_P + |\downarrow_x\rangle_P)$), the particles in this experiment do not determinately follow path A nor do they determinately follow path B; rather, the standard theory tells us that they follow a superposition of the two paths (which is what ultimately explains the interference statistics at I). Yet whenever we look for a particle, its state instantaneously and randomly jumps to an eigenstate of position. Since the position of each particle is correlated with its x-spin (this is all the x-spin sorting device does), once a particle has a determinate position (on path A or path B), it also has a determinate x-spin (up or down, respectively), it behaves accordingly, and thus changes the statistics at I to those one would expect from an electron with a determinate x-spin.

It is the fact that a particle in an eigenstate of z-spin is in a superposition of x-spin states and that a particle in an eigenstate of x-spin is in a superposition of z-spin states that explains, in the standard theory, why we cannot simultaneously know both the z-spin and the x-spin of a particle. It is not that the particle has both properties simultaneously and that we just cannot know what they are; rather, it is that a particle with a determinate z-spin simply has no determinate x-spin and a particle with a determinate x-spin simply has no determinate z-spin.

1.3 *A more exotic example: The curious behaviour of neutral* K *mesons*

As suggested earlier, a particle can be in a superposition of most any particle property. It is even possible for a particle to be in a superposition of being two fundamentally different types of particles. Neutral K mesons provide an example of this crazy sort of behaviour.

Neutral K mesons are produced in strong interaction processes like

$$\pi^- + p \rightarrow \Lambda^0 + K^0. \tag{1.1}$$

Charge is conserved in this interaction (as it is in all known interactions): the π meson is negatively charged, the proton is positively charged, and the Λ^0 particle and K^0 meson are both neutral. Strangeness, another basic property of matter, is also conserved here (as it is in all strong interactions): both particles on the left of the reaction have strangeness zero, the Λ^0 particle has strangeness -1 and the K^0 meson has strangeness $+1$.

Many particles have corresponding antiparticles with different physical properties. If a particle is positively charged, its antiparticle would be negatively charged. If a particle has strangeness $+1$, its antiparticle would have a strangeness -1. And so on. And a particle and its antiparticle would typically destroy each other if they ever met. The K^0 meson's antiparticle is the $\bar{K}^0$ meson, which has no charge and has a strangeness -1. A curious thing about neutral K mesons (and there are many curious things about these particles[4]) is that they are typically found to be in superpositions of being K^0 mesons and being $\bar{K}^0$ mesons *at the same time*. That is, on the standard interpretation of quantum-mechanical states, a neutral K meson is typically not a K^0 meson, not a $\bar{K}^0$ meson, not both, and not neither; rather, it is in a superposition of being both types of particle simultaneously.[5]

One might represent the state where a neutral K meson is determinately a K^0 particle as $|K^0\rangle$ and the state where it is determinately a $\bar{K}^0$ particle as $|\bar{K}^0\rangle$. One might then represent the states of two other types of neutral K mesons, K_S and K_L, as linear superpositions of determinate K^0 and $\bar{K}^0$ states:

$$|K_S\rangle = \frac{1}{\sqrt{2}}(|K^0\rangle + |\bar{K}^0\rangle) \tag{1.2}$$

and

$$|K_L\rangle = \frac{1}{\sqrt{2}}(|K^0\rangle - |\bar{K}^0\rangle). \tag{1.3}$$

This is the same notation used above to represent a particle in a superposition of different x-spins—in fact, the mathematical representation of neutral K meson states is exactly the same as the representation I shall use for x-spin and z-spin states throughout the book. I shall discuss how to represent physical states in quantum mechanics further in the next chapter. For now all that matters is that one recognizes that $|K_S\rangle$ and $|K_L\rangle$ are *different* superpositions of a determinate K^0 and a determinate $\bar{K}^0$ state.

---

[4] They are, for example, involved in interactions that suggest that time-reversal symmetry and parity must be violated by our most basic physical laws.

[5] A similar sort of superposition is quite common in relativistic quantum field theory. Indeed, a field is often thought of as a superposition of different particle configurations, each configuration containing a different number of fundamental particles in different positions. When one observes the particle configuration, on the standard collapse story, the state instantaneously evolves to one of the superposed configurations.

K_S and K_L mesons both decay via weak interactions, but in very different ways. K_S mesons decay as follows:

$$K_S \rightarrow \pi^+ + \pi^- \text{ or } K_S \rightarrow \pi^0 + \pi^0 \tag{1.4}$$

and these processes occur in a time $\tau_S \approx 0.9 \times 10^{-10}$ sec. K_L mesons decay as follows:

$$K_L \rightarrow \pi\pi\pi, \, K_L \rightarrow \pi e v, \text{ or } K_L \rightarrow \pi \mu v \tag{1.5}$$

and these processes occur in a time that is over 500 times as long as the K_S mesons, $\tau_L \approx 518 \times 10^{-10}$ sec. So the K_S and K_L are empirically distinguishable. Indeed, since the different types of K mesons have very different physical properties, they really do seem to be fundamentally different fundamental particles.

The state of a neutral K meson evolves even when the particle is left alone. On the standard collapse theory, what happens is that a K^0 meson would evolve to a neutral K meson in a superposition of being a $\bar{K}^0$ meson and being a K^0 meson, and in this state it has a nonzero probability of collapsing to a $\bar{K}^0$ meson if we checked to see what sort of K meson it is. Consequently, it is possible to start with a K^0 meson, then find it to be its own antiparticle, a $\bar{K}^0$ meson, when it is observed later. There is a sense in which one might say that such a particle would not even have an identity! Figure 1.5 shows the probability of observing a $\bar{K}^0$ meson at time t if the state at time t_0 is K^0. Time is represented in units of τ_S (see Baym 1969: 42–3 for the explicit expression of the dynamics). Such theoretical calculations agree well with the results of actual experiments.[6]

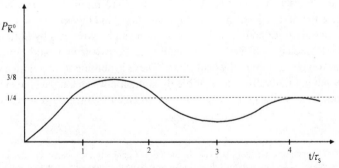

FIG. 1.5 Probability of observing $\bar{K}^0$ at t if the particle is $\bar{K}^0$ at t_0.

[6] Recent experiments suggest that certain types of neutrinos may exhibit a similar sort of oscillating behaviour (and consequently have a nonzero rest mass).

1.4   *The measurement problem*

In order to account for the crazy wave-like behaviour of matter yet ensure that we typically get determinate results to our measurements, the standard von Neumann–Dirac theory has two dynamical laws: (1) the linear wave equation describes the time-evolution of all unobserved systems and accounts for their wave-like behaviour[7] and (2) the random collapse dynamics describes what happens whenever an observation is made and accounts for the fact that we always find systems to have determinate properties whenever we observe them. If the standard theory is right, then according to these laws, the crazy quantum-mechanical behaviour is not restricted to *microscopic* systems—it is just restricted to *unobserved* systems. If the theory is right, then chairs are typically in superpositions of different locations, cats are typically in superpositions of being alive and dead, etc. *as long as no one is watching*. Of course, one can describe situations where the wave function associated with a macroscopic system is as well-localized as one wants, but as John Bell once put it, one can also describe situations where the wave function associated with such a system is as spread-out as one does *not* want. This wave-like behaviour of even macroscopic objects is counter-intuitive, but there is nothing inherently wrong with a counter-intuitive theory—after all, given the odd behaviour of matter, any empirically adequate formulation of quantum mechanics will be at least to some extent counter-intuitive. Again, the quantum measurement problem is not that quantum mechanics is counter-intuitive—it is that the theory is at least ambiguous and perhaps logically inconsistent.

In the context of the standard theory, the measurement problem results from the fact that the two dynamical laws are mutually incompatible. Since the first is deterministic and continuous and the second is stochastic and discontinuous, no physical system can be governed by both laws simultaneously—indeed, as we shall see, the two laws would typically lead to very different physical states. There is nothing wrong with a theory having mutually incompatible dynamical laws as long as it also provides clear and disjoint conditions for when each correctly describes the

---

[7] An operator on vectors is linear if and only if applying the operator to the sum of two vectors yields the same result as applying it to each of the vectors individually, then adding each of these results. The deterministic dynamics (the dynamics described by the time-dependent Schrödinger wave equation) is linear because it can be represented by a family of linear operators each of which takes the state of the system at an initial time to the state of the system at some other time.

evolution of a system, but this is where our loose talk of things behaving one way when someone is looking and another way when no one is looking catches up to us. The standard theory tells us that the deterministic dynamics describes the evolution of a system unless it is *measured*, in which case the random dynamics kicks in. But the theory does not tell us what constitutes a *measurement*. One is left to one's own intuitions concerning what interactions ought to count as measurements. While it turns out that one can typically use such intuitions to get good empirical predictions from the theory, the fact that our intuitions concerning what it takes for an interaction to count as a measurement are ultimately vague means that quantum mechanics is at best ambiguous. Further, if one supposes that measuring devices are ordinary physical systems just like any other, constructed of fundamental particles interacting in their usual deterministic way (and why wouldn't they be?), then the standard theory is logically inconsistent since no system can obey both the deterministic and stochastic dynamical laws simultaneously. This is the measurement problem.

The long and continued empirical success of the standard theory makes it possible for most working physicists simply to ignore the measurement problem. Indeed, perhaps the most popular position right now is that the foundational problems of quantum mechanics were solved long ago, that Niels Bohr showed that criticisms of quantum mechanics in general and Albert Einstein's criticisms in particular were naive and misguided, that there was in fact nothing wrong with the theory. Bohr was so successful in his defence of quantum mechanics that for many years criticism of the theory was taken to imply incompetence.[8] Recently, however, Bohr's position has lost some of its hold on the physics community. At the 1976 Nobel Conference Murray Gell-Mann lamented that

The fact that an adequate philosophical presentation [of quantum mechanics] has been so long delayed is no doubt caused by the fact that Niels Bohr brainwashed

---

[8] Jeremy Bernstein recounts an incident involving Robert Oppenheimer that illustrates the degree of consensus eventually generated by Bohr: 'I once saw Oppenheimer reduce a young physicist nearly to tears by telling him a talk he was delivering on the quantum theory of measurement at the Institute was of no interest, since all the problems had been solved by Bohr and his associates two decades earlier' (Bernstein 1991: 63). It is curious that Bohr and his Copenhagen interpretation of quantum mechanics could have led to such consensus when there is good reason to suppose that few physicists ever really understood what Bohr's position was. Indeed, it is difficult to find a single, unambiguous statement anywhere in the literature of what the Copenhagen interpretation was. It was the von Neumann–Dirac formulation that made it into the textbooks, and so this was presumably the formulation that in fact held the allegiance of most physicists.

a whole generation of theorists into thinking that the job was done fifty years ago. (Gell-Mann 1979: 29)

And some years after winning the Nobel Prize, Willis Lamb wrote:

A discussion of the interpretation of quantum mechanics on any level beyond this almost inevitably becomes rather vague. The major difficulty involves the concept of 'measurement' ... I have taught graduate courses in quantum mechanics for over 20 years at Columbia, Stanford, Oxford and Yale, and for almost all of them have dealt with measurement in the following manner. On beginning the lectures I told the students, 'You must first learn the rules of calculation in quantum mechanics, and then I will tell you about the theory of measurement and discuss the meaning of the subject.' Almost invariably, the time allotted to the course ran out before I had to fulfill my promise. (1969; quoted in Wheeler and Zurek 1983, pp. xviii–xix)

Albert Einstein, Erwin Schrödinger, Eugene Wigner, David Bohm, John Bell, and others have been influential over the years in arguing against the standard formulation of quantum mechanics. Indeed, Bell apparently shared many of Einstein's intuitions concerning what a satisfactory version of quantum mechanics would look like.[9]

The problem again is not that the standard theory fails to make the right empirical predictions. When supplemented with the right intuitions, quantum mechanics makes remarkably accurate empirical predictions for the sort of experiments that we in fact perform. And the problem is not that the standard formulation of quantum mechanics is counter-intuitive either. While it would be silly to seek a counter-intuitive theory for the sake of novelty, we know that any empirically adequate theory must be counter-intuitive in order to account for the quantum weirdness we routinely observe in small and well-isolated systems. Rather, the problem is that the standard theory as it stands cannot be taken as providing a complete and accurate description of the behaviour of the systems we observe and the devices we use to make these observations.

This is where Hugh Everett III comes in. Everett's proposal for solving the measurement problem was simply to drop the collapse dynamics from the standard formulation of quantum mechanics, take the resulting theory as providing a complete and accurate description of all physical processes without exception, then deduce the standard predictions of quantum mechanics as subjective appearances to observers treated as

---

[9] Bell argued that 'Quantum Mechanics is, at best, incomplete' (1987:26). See Bell (1987: 81–92) for his own account of Einstein's intuitions.

systems within the revised theory (Everett 1973: 9–10, 1957*a*: 2, 1957*b*: 315). The first step of this proposal would clearly eliminate any possible conflict between the two dynamical laws, and in this sense it would clearly solve the measurement problem. But while Everett's proposal may sound relatively straightforward, it turns out that there is little consensus on the details of how it was supposed to work. Indeed, there are probably few topics in the history of twentieth-century physics that have engendered more debate and confusion than Everett's interpretation of quantum mechanics.[10] Further, since it is the collapse dynamics that ensures that we end up with determinate records at the end of a measurement in the standard theory, if we drop the collapse dynamics, then we have to find a new explanation for the determinateness of our experience.[11]

I do not intend to give an exhaustive discussion of everything everyone has said about how Everett ought to be interpreted. Rather, I shall describe what Everett himself said in considerable detail, then focus on a few of the more influential criticisms and reconstructions of his position and compare and contrast some of the problems and virtues of each of these. Throughout I shall try to provide ample evidence of what people have actually said. If nothing else, I hope that this will help to sharpen the various debates involving Everett's relative-state theory by pointing out some of the ambiguities that must be resolved before a sensible discussion can even get started. But it would be nice if we could also gain some insight into how one might solve the quantum measurement problem.

[10] The mere mention of Everett's interpretation in a group of physicists or philosophers of science can invoke a long discussion filled with little agreement. As Abner Shimony once complained, 'I briefly mentioned [Everett's interpretation] in my original lecture, and then most of the discussion following the lecture was devoted to it. This is not the first time I have observed a discussion of the Everett interpretation expanding to fill the available vessel' (Shimony 1986: 201).

[11] As we shall see, when and whether a collapse occurs at least in principle has empirical consequences. This means that the measurement problem is a problem even for those who are instrumentalists with respect to quantum mechanics if they want a theory that provides coherent empirical predictions in all physical situations describable in the language of the theory.

2

THE STANDARD FORMULATION OF QUANTUM MECHANICS

H UGH E VERETT III presented his relative-state formulation of quantum mechanics as a new, more general and complete theory than the standard von Neumann–Dirac theory, one that entailed the same subjective appearances but avoided the measurement problem. In order to understand what Everett might have had in mind, one must first understand how the standard formulation of quantum mechanics works and the problems it encounters.

2.1 *The foundations of a new theory*

Classical mechanics, the theory of motion described by Isaac Newton in the 1600s, faced no serious empirical challenges for more than 200 years, but in the late 1800s people began to notice phenomena that classical mechanics could not explain. Rather than completely abandon classical mechanics, physicists, led by Niels Bohr, began by developing rules of thumb for when microscopic systems could be treated classically and when they could not and rules of thumb for how to make empirical predictions when a system was not behaving classically. But it was not until the mid-1920s that a truly systematic theory of quantum mechanics began to be developed.

The first description of the basic principles of the new quantum mechanics was given by Werner Heisenberg in 1925. In response to Bohr's old rules of thumb, Heisenberg argued: 'It is better . . . to admit that the partial agreement of the quantum rules with experiment is more or less accidental, and try to develop a quantum theoretical mechanics, analogous to the classical mechanics in which only relations between observable quantities appear' (Heisenberg 1925; trans. and quoted in Pais 1988: 253). Heisenberg's theory came to be known as *matrix mechanics*, and many of the details of the theory were filled in within the year by Werner Heisenberg, Wolfgang Pauli, Max Born, E. P. Jordan, and P. A. M. Dirac.

Schrödinger described an alternative approach for formulating a general theory of quantum mechanics (Schrödinger 1926a). His theory was based on de Broglie's suggestion that the relationship between photons (light-quanta) and electromagnetic wave phenomena that was postulated to explain the photoelectric effect ought to be generalized to include all material particles and that all particles should thus be expected to exhibit wave-like properties. Schrödinger's theory came to be known as *wave mechanics*.

Schrödinger's (1926b) paper containing the one-particle, nonrelativistic, time-dependent wave equation was received for publication in June 1926. Schrödinger's wave equation is:

$$i\hbar \frac{\partial}{\partial t} \psi(x, t) = \hat{H} \psi(x, t), \qquad (2.1)$$

where $i = \sqrt{-1}, \hbar = h/2\pi$ (where h is Planck's constant), $\psi(x, t)$ is the wave function, and $\hat{H}$ is the Hamiltonian of the system, which is determined by the system's energy properties. This equation gives a dynamics for the evolution of the wave function, which Schrödinger initially understood as a matter-wave. This evolution is linear and perfectly deterministic: if one knows the initial wave function $\psi(x, 0)$ and the Hamiltonian $\hat{H}$, then one can in principle predict what $\psi(x, t)$ will be at any time in the future or past.

Schrödinger thought that the wave function associated with an electron somehow represented the mass distribution and charge density of the electron. More generally, he thought that waves, rather than particles, were the most basic constituents of matter. Along these lines, he thought that the particle-like behaviour of matter could be accounted for by the motions of narrow, coherent wave packets. Schrödinger convinced himself that this was possible by considering the motion of the wave packet representing an electron in a harmonic potential. Since he was able to show that the wave packet did not disperse in a harmonic potential, he thought that this explained the particle-like behaviour of electrons. Schrödinger believed that the deterministic wave equation provided a complete and accurate description of the time-evolution of all physical systems and that the objects of quantum mechanics were no more mysterious than classical waves (Pais 1988: 256).

But Schrödinger's theory soon encountered problems. It was almost immediately pointed out that the harmonic oscillator was the exception rather than the rule and that the wave packet representing an electron would typically disperse very quickly on Schrödinger's deterministic dynamics. This, of course, undermined Schrödinger's interpretation of the wave function since it would presumably be impossible to derive particle-like properties from spread-out waves. But it also suggested that Schrödinger was wrong to claim that one can get by with a deterministic formulation of quantum mechanics based solely on his deterministic dynamics. His dynamics could not be universally valid since we do in fact observe localized particles in situations where his dynamics alone would typically predict spread-out waves.

Four days after Schrödinger's June paper, a paper by Max Born was received for publication (Born 1926*a*). In this paper Born proposed keeping Schrödinger's mathematical formalism and his deterministic dynamics (though now with a new role), but he wanted to drop Schrödinger's interpretation of the quantum-mechanical waves altogether. Rather than being the fundamental objects that constitute the real world, Born took Schrödinger's waves to be descriptive of particles, which were then taken to be fundamental. Born proposed that a particle's wave function determines the probabilities of finding that particle at various positions if one looks for it.[1] The idea was to interpret the wave intensity not as the density of an actual distribution of matter as Schrödinger had wanted but as a probability density for the presence of a particle.

Born's next paper was received in July of the same year (Born 1926*b*). In this paper he notes that whenever the state of a system ψ can be expressed as a sum of determinate states ψ_n, each with a determinate energy

$$\psi = \sum_n c_n \psi_n, \tag{2.2}$$

then $|c_n|^2$ represents the probability of the system being found to be in the determinate energy state ψ_n. As is the case with position, then, quantum mechanics typically does not describe a system as having a particular determinate energy. Rather, it tells us what the probability is for finding a system to have a particular energy. And when the state is written as a sum of determinate energy states, then the probability of each state is given by the norm-squared of the coefficient on the term describing the system as determinately having that energy.

Born also suggested that his statistical interpretation of the wave function required a new dynamical law, that the deterministic wave equation

---

[1] He added in a footnote that further thought had revealed that the probabilities were actually proportional to the *square* of the wave function (Pais 1988: 256–7).

provided only part of the quantum dynamics. Born explained that 'the motion of particles follows probability laws but the probability itself propagates according to the law of causality'.[2] That is, while the state transitions that accounted for the experimental results were acausal (indeterministic) processes about which one could only have probabilistic laws, the evolution of these probabilities was correctly described by the causal (deterministic) Schrödinger dynamics. This was perhaps the first explicit suggestion that quantum mechanics required two dynamical laws.

The physical picture that Born ended up with, then, was one where forces, by way of Schrödinger's linear dynamics, determined the evolution of probabilities, and where these probabilities in turn described the random evolution of states.[3] He further explained that, unlike the probabilities that arise in classical thermodynamics, the probabilities of quantum mechanics were not simply a matter of our lack of knowledge, but rather they were the result of fundamentally random, acausal processes and consequently could not be eliminated from the theory (Pais 1988: 258). Heisenberg agreed and concluded that 'quantum mechanics establishes the final failure of causality'.[4]

Born did not say exactly how or when random state transitions occurred. Indeed, he felt that there was an element of mystery as to what physical events were involved in a state transition that made it difficult to talk about and perhaps impossible to picture: 'Whatever happens during the transition can hardly be described within the conceptual framework of Bohr's [old quantum theory], nay, probably in no language which lends itself to visualizability.'[5] For his part, Bohr agreed and did his best to add to the mystery. He explained that quantum mechanics requires one to renounce 'the causal space-time co-ordination of atomic processes'.[6] Along these lines, Bohr argued that since there are no absolute matters of fact about such things (only 'complementary descriptions'), one must for ever give up the possibility of providing causal explanations for when, where, or

---

[2] Born said: 'Die Beweggung der Partikel folgt Wahrscheinlichkeitsgesetzen, die Wahrscheinlichkeit selbst aber breitet sich im Einklang mit dem Kausalgesetz aus' (Born 1926: 804).

[3] 'We free forces of their classical duty of determining directly the motion of particles and allow them instead to determine the probability of states' (quoted in Jammer 1974: 305).

[4] Heisenberg (1927); trans. and repr. in Wheeler and Zurek (1983: 83). The word *acausal* usually means *indeterministic* in these early quotations.

[5] Quoted in Jammer (1974: 306).

[6] Bohr's Como lecture, 1927.

why physical events occur at the atomic level. Instead of providing causal descriptions of how physical systems evolve over time, he felt that the proper function of a physical theory was to provide empirical predictions for what one should expect to observe under *classically* specified experimental conditions. And one can certainly see the appeal of trying to avoid telling a detailed causal story here.

On what came to be known as the Copenhagen interpretation, quantum mechanics had no responsibility to provide a fine-grained account of what happens and why, so one is never put in the embarrassing position of trying to explain why the usual deterministic evolution of the state is suspended during the random state transitions. Rather, quantum mechanics is to be judged by how well it serves as an algorithm for making empirical predictions in the context of experiments where the state of the measurement apparatus can be described classically. In response to the complaint that quantum mechanics did not seem to provide a complete description of physical processes, Heisenberg said with confidence that 'the presumption that behind the perceived statistical world there still hides a "real" world in which causality holds' is 'fruitless and senseless' since 'physics ought to describe only the correlation of observations' (Wheeler and Zurek 1983: 83).

Few physicists, however, could consistently resist the temptation of thinking that quantum mechanics ought to provide a precise description of the time-evolution of the complete physical state of a system under all circumstances. What was sacrificed instead was the belief that a complete description of a physical system would determine all of its classical properties. More specifically, the orthodox position became that the quantum-mechanical state provided a complete and accurate physical description of a system even though it typically did not determine the values of all the classical physical quantities (quantities like position, momentum, and energy). But having made this sacrifice, physicists were still interested in saying precisely how the quantum-mechanical state evolved.

2.2   *The collapse of the wave function*

Einstein did not like Born's formulation of quantum mechanics. Within five months of the publication of the papers where Born described his new statistical interpretation of the wave function, Einstein had rejected it. Einstein told Born that his formulation of quantum mechanics was 'certainly imposing ... but an inner voice tells me that it is not yet the real

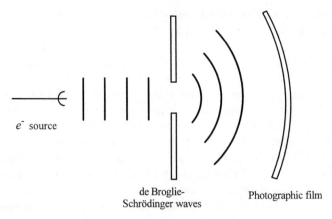

e^- source

de Broglie-
Schrödinger waves

Photographic film

FIG. 2.1 Einstein's Solvay experiment.

thing' (Einstein 1926: 91). Born later reported that

Einstein's verdict on quantum mechanics came as a hard blow to me. He rejected it not for any definite reason, but rather referred to an 'inner voice.' ... [This rejection] was based on a basic difference of philosophical attitude, which separated Einstein from the younger generation to which I felt that I belonged, although I was only a few years younger than Einstein. (Born 1971: 91)

At the 1927 Solvay Congress Einstein tried to explain what it was that he did not like about quantum mechanics. He presented a thought experiment where electrons are sent through an aperture, travel some distance, then hit a large semispherical sheet of photographic film (Fig. 2.1). Because of the size of the aperture and the velocity of the particles, Einstein argued that the de Broglie–Schrödinger wave associated with each electron would be diffracted in such a way that it would hit more or less the entire surface of the film simultaneously. And he considered two ways that one might understand these waves: (I) one might suppose that they describe a cloud of electrons spread through space, and that quantum mechanics, consequently, does not describe individual processes but only an ensemble of an infinity of elementary processes or (II) one might suppose that quantum mechanics is meant to provide a complete description of the individual physical processes, where each electron is somehow individually described by the wave packet, these waves diffract, then the particle somehow ends up hitting a single point on the film even though the wave itself strikes the entire surface of the screen.

Einstein rejected (I) because he felt that it was incompatible with experimental results,[7] but he did not like (II) either. On this view, the electron goes from a spread-out state to a localized state instantaneously—the electron wave is spread across the screen at one instant and a moment later is concentrated at a single point, the point where we find the electron. And Einstein thought that this required a single elementary process to produce an action at two or more places on the screen simultaneously, and this, he thought, required one to postulate some mechanism for action at a distance—a mechanism that would send all of the quantum-mechanical wave amplitude to zero at all but one point simultaneously (Instituts Solvay 1928: 255). Einstein believed that taking the wave function as a complete and accurate description of the state of a single electron was incompatible with relativity: 'If one works solely with the Schrödinger waves, interpretation (II) of $|\psi|^2$ implies, to my mind, a contradiction with the postulate of relativity.'[8] I believe that this worry about the compatibility of quantum mechanics and relativity formed the basis for Einstein's lifelong mistrust of quantum mechanics. Most of his colleagues thought that his worries about a potential conflict were unfounded, but as we shall see, while the usual linear dynamics and the statistical predictions of quantum mechanics can be made compatible with relativity, when one tries to solve the measurement problem, one often ends up postulating an auxiliary dynamical law that cannot.

Max Jammer has argued that in response to Einstein's worries, Heisenberg, Pauli, and Dirac took the position that the wave function does not represent a course of events in space and time, but rather, represents *our knowledge* about such events (1974: 373). While I do not entirely agree with this interpretation of the Solvay discussion,[9] there is certainly a long tradition of interpreting the quantum-mechanical state as representing our state of knowledge. The short story is that it does not work, at least not in any straightforward way.

The longer story goes something like this. Suppose that, contrary to the standard interpretation of states, a system S is as a matter of fact

---

[7] Einstein mentioned the Geiger and Bothe experiments (Instituts Solvay 1928: 255).

[8] 'Si l'on opère uniquement avec les ondes de Schrödinger, l'interpretation II de $|\psi|^2$ implique à mon sens une contradiction avec le postulat de la relativité' (Instituts Solvay 1928: 256).

[9] See e.g. what Dirac says about the process of collapse; further, the interpretation of the wave function that Jammer ascribes to Heisenberg, Pauli, and Dirac seems to be incompatible with the principle of state completeness, which was something that I believe all three held by this time.

either in box A or in box B and that its quantum-mechanical state $1/\sqrt{2}(|A\rangle_S + |B\rangle_S)$ simply represents our lack of knowledge as to which box actually contains the system. In this case, when we find the system in box A, the probability of it being in box B would go from 1/2 to 0, but this would correspond to a change in the subjective beliefs of an observer and not to any physical events at A or B since the system was in box A all along. All physical events would be perfectly local on this story, and there would be no contradiction with relativity. Consequently, this sort of straightforward epistemic interpretation of the quantum-mechanical state would be immune to Einstein's criticism. But if the quantum-mechanical state of S is $1/\sqrt{2}(|A\rangle_S + |B\rangle_S)$, it cannot be that S is either determinately in box A or determinately in box B. The reason is that a system in a superposition of $|A\rangle_S$ and $|B\rangle_S$ has physical dispositions that are quite different from those it would have in state $|A\rangle_S$ or in state $|B\rangle_S$. Consider an observable O that has $1/\sqrt{2}(|A\rangle_S + |B\rangle_S)$ as an eigenstate with eigenvalue $+1$ and every orthogonal state as an eigenstate with eigenvalue -1. If S were in the state $1/\sqrt{2}(|A\rangle_S + |B\rangle_S)$, then a measurement of O would give the result $+1$ with certainty, but if S were in the state $|A\rangle_S$ or $|B\rangle_S$, then a measurement of O would give the result $+1$ with only probability 1/2 (and it seems to me that there is good historical evidence that Dirac and others knew very early on that the ignorance interpretation of quantum-mechanical states did not work because it failed to explain such interference effects).

After a short discussion of some of Einstein's worries about relativity by Pauli and Lorentz, Dirac offered a description of how he understood quantum transitions and the measurement process:

This theory describes the state of the world at any given moment by a wave function ψ, which normally varies according to a causal law in such a way that its initial value determines its value at any later moment. It may happen, however, that at a given moment t_1, ψ may be expanded into a series of the form $\psi = \sum_n c_n \psi_n$ in which the ψ are wave functions which cannot interfere with each other at times later than t_1. In this case, the state of the world at times later than t_1 will be described not by ψ, but by one of the ψ_n. One can say that nature chooses the particular ψ_n that is suitable, since the only information given by the theory is that the probability that any one of the ψ_n will be selected is $|c_n|^2$. Once made, the choice is irrevocable and will affect the entire future state of the world. The value of n chosen by nature can be determined by experiment and *the results of all experiments* are numbers that describe such choices of nature. (Instituts Solvay 1928: 262)

Note that while Dirac seems to take the reduction of the wave packet to be a physically real process, he does not say that the choice is made by nature when a measurement is made. Rather, he says that nature irrevocably picks one of the components of the wave function to represent the state of the world *when the wave function's components are no longer able to interfere with each other*. This condition foreshadows recent decoherence formulations of quantum mechanics. I shall discuss precisely what decoherence is, how it might be used to address the measurement problem, and some of the problems with this approach later (Chapter 8), but even here one might be a little puzzled by Dirac's proposal. First, the condition for *when* nature is supposed to make its irrevocable choice is unclear. Dirac says that this choice occurs when the terms of the superposition are no longer able to interfere with each other; but this never really happens— there is always, at least in principle, *some* experiment that would detect interference effects between terms. Secondly, Dirac's proposal is ambiguous concerning what sort of state nature chooses. Nature is supposed to choose a particular term in some expansion of the wave function, but how does nature decide which expansion to use? After all, there is a continuous infinity of different expansions from which to choose the new state.

Heisenberg strongly disagreed with Dirac's claim that nature chooses the component of the wave function that will represent the state of the world. His disagreement was related to the question when Dirac thought that nature was supposed to make its choice. Heisenberg argued that one could perform experiments where one sees interference effects lasting for very long times, but that if nature makes the choice, it would be difficult to explain how such interference effects were produced. Rather than nature, Heisenberg insisted that it was the observer who makes the choice because it is only when an observation is made that the 'choice' becomes a physical reality and the interference effects are destroyed (Instituts Solvay 1928: 264–5). But, of course, if an observer could really choose which state his object system collapsed to, then one would be able to perform experiments where an observer would cause a system to violate the standard statistical predictions of quantum mechanics (which would be remarkable indeed). So when Heisenberg said that it was the observer who made the choice, he must have meant that it was an observer's act of observation that then forced nature to make a particular choice. That is, the observer decides when a collapse will occur and to what *set* of eigenstates by deciding when to make a measurement and what observable to measure; it is then nature that randomly and irrevocably chooses one eigenstate as actual from the set of eigenstates determined by the observer's choice—a choice is made by nature only in so far as the observer does not decide which eigenstate of the observable being measured ends up providing a complete and accurate description of the state of the system after the measurement.

By 1930 Dirac's position had moved closer to Heisenberg's in that he now explicitly recognized the role of observation in the dynamics. Dirac noted that while Schrödinger's deterministic dynamics applies to almost all interactions, 'There are, however, two cases when we are in general obliged to consider the disturbance as causing a change in the state of the system, namely, when the disturbance is an observation and when it consists in preparing the system so as to be in a given state' (Dirac 1930: 9). In these cases, the time-evolution of the state is no longer correctly described by the deterministic dynamics.

The introduction of indeterminacy into the results of observations, which we had to make in our discussion of the photon, must now be extended to the general case. When an observation is made on an atomic system that has been prepared in a given way and is thus in a given state, the result will not in general be determinate; that is, if the experiment is repeated several times under identical conditions, several different results may be obtained.[10] (10)

His argument for the necessity of a different dynamics for observations appeals to what it means for a system to be in a superposition: 'When a state is formed by the superposition of two other states, it will have properties that are in a certain way intermediate between those of the two original states and that approach more or less closely to those of either of them according to the greater or less "weight" attached to this state in the superposition process' (8). It is the sense in which a superposition of two states is intermediate between the two states that requires stochastic indeterminacy in the results of measurements.

The indeterminacy in the results of observations is a necessary consequence of the superposition relationships that quantum mechanics requires to exist between the states. Suppose that we have two states A and B such that there exists an observation which, when made on the system in state A, is certain to lead to one particular result, and when made on the system in state B is certain not to lead to this result. Two such states we call *orthogonal*. Suppose now that the observation is made on the system in a state formed by a superposition of A and B. It is impossible for the result still to be determinate (except in the special case when the weight of A or B in the superposition process is zero). There must be a finite

---

[10] Note that Dirac's use of the term *determinate* is very different from my use of the same term throughout the text.

probability *p* that the result, that was certain for state *A*, will now be obtained and a finite probability 1 − *p* that it will not be obtained. By continuously varying the relative weights in the superposition process we can get a continuous range of states, extending from pure *A* to pure *B*, for which the probability of the result, that was certain for state *A*, being obtained varies continuously from unity to zero. (10)

Dirac also recognized that, because of the repeatability of measurement results, the result of a measurement must determine the state of a system after the measurement.

Consider an observation, consisting of the measurement of an observable *α*, to be made on a system in the state *ψ*. The state of the system after the observation must be an eigenstate of *α*, since the result of a measurement of *α* for this state must be a certainty. (Dirac 1930: 49)

So the probability of a given result being obtained when an observation is made on a system in a given state must be the probability of the state of the system *collapsing* to the corresponding eigenstate over the course of the observation.

Dirac's main argument for the collapse dynamics then went as follows: If the quantum-mechanical state of a system is not an eigenstate of the physical quantity we intend to measure, then Born's statistical interpretation tells us that we can only know what the probabilities are of it being found to have a particular value. But if we make the measurement and find out what the value of the physical quantity is, then, since we now know what the value is and since we can repeat the measurement to see that we do indeed know what the value is, the system must now be in a state that describes it as having that value with probability one. But according to Born's statistical interpretation, there is only one state that has this property, and it is the eigenstate of the observable that corresponds to the result we got. Schrödinger's dynamics, however, will typically not predict anything even close to this as the post-measurement state. So the evolution of the system must in fact violate Schrödinger's dynamics by randomly jumping to an eigenstate of the observable we are measuring.

Suppose that we build an *x*-spin measuring device *M* so that when a system *S* starts in a determinately *x*-spin up state and *M* starts in a ready-to-make-a-measurement state, the *x*-spin of *S* is undisturbed and *M* ends up reporting that the measurement result was *up* and that when *S* starts in an *x*-spin down eigenstate and *M* starts in a ready-to-make-a-measurement state, the *x*-spin of *S* is undisturbed and *M* ends up reporting that the measurement result was *down* (Fig. 2.2). That is, suppose that *M*

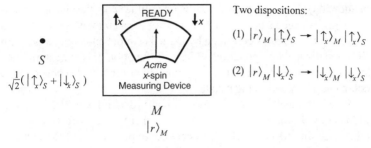

FIG. 2.2 The x-spin setup.

has the following two dispositions:

(1) $|\text{ready}\rangle_M |\uparrow\rangle_S \longrightarrow |\text{up}\rangle_M |\uparrow\rangle_S$

and

(2) $|\text{ready}\rangle_M |\downarrow\rangle_S \longrightarrow |\text{down}\rangle_M |\downarrow\rangle_S$.

Now suppose that M measures the x-spin of S when it is initially in an eigenstate of z-spin, which is a superposition of x-spin up and x-spin down states, and suppose that the interaction between the two systems is described by the linear dynamics. The composite system $M + S$ would begin in the state

$$|\text{ready}\rangle_M 1/\sqrt{2}(|\uparrow\rangle_S + |\downarrow\rangle_S) = 1/\sqrt{2}(|\text{ready}\rangle_M |\uparrow\rangle_S + |\text{ready}\rangle_M |\downarrow\rangle_S). \tag{2.3}$$

Since the usual Schrödinger dynamics is linear and since our measuring device has dispositions (1) and (2) above, the composite system will evolve to the state

$$1/\sqrt{2}(|\text{up}\rangle_M |\uparrow\rangle_S + |\text{down}\rangle_M |\downarrow\rangle_S), \tag{2.4}$$

which is a superposition of M recording x-spin up and S being x-spin up and M recording x-spin down and S being x-spin down (because the dynamics is linear, one can figure out what happens to each term of the initial linear superposition individually, then take the linear superposition of these as the final state). The problem is that if one takes this to be a complete and accurate description of the post-measurement state, then one cannot say that M recorded a determinate result; or more precisely, one cannot say that M recorded *up* and one cannot say that M recorded *down*. In order to take the quantum-mechanical state as complete and accurate and to account for the repeatability of measurements, one must suppose that the post-measurement state is either $|\text{up}\rangle_M |\uparrow\rangle_S$

or $|down\rangle_M|\downarrow\rangle_S$, which are each incompatible with the linear dynamics. Thus, we need a new dynamical law for measurement processes, and the collapse dynamics described by Dirac is simple and compatible with Born's statistical interpretation of the wave function. It is our determinate experience of one result or the other that tells us that (2.4) cannot be the complete post-measurement state.

Dirac continued to argue for the collapse dynamics in later editions of his book. In the 1957 edition he argued that the standard eigenvalue–eigenstate link together with 'physical continuity' entailed the collapse postulate. The eigenvalue–eigenstate link, which provides a way of understanding the relationship between Born's statistical interpretation and the doctrine of wave-function completeness, says that

If the dynamical system is in an eigenstate of a real dynamical variable ξ, belonging to the eigenvalue ξ', then a measurement of ξ will certainly give as a result the number ξ'. Conversely, if the system is in a state such that a measurement of a real dynamical variable ξ is certain to give one particular result ... then the state is an eigenstate of ξ and the result of the measurement is the eigenvalue of ξ to which this eigenstate belongs. (Dirac 1958: 35)

And from this Dirac argued that there must be a collapse of the state on measurement:

When we measure a real dynamical variable ξ, the disturbance involved in the act of measurement causes a jump in the state of the dynamical system. From physical continuity, if we make a second measurement of the same dynamical variable ξ immediately after the first, the result of the second measurement must be the same as the first. Thus after the first measurement has been made, there is no indeterminacy in the result of the second. Hence after the first measurement is made, the system is in an eigenstate of the dynamical variable ξ, the eigenvalue it belongs to being equal to the result of the first measurement. This conclusion must still hold if the second measurement is not actually made. In this way we see that a measurement always causes the system to jump into an eigenstate of the dynamical variable that is being measured, the eigenvalue this eigenstate belongs to being equal to the result of the measurement. (1958: 36)

And thus we have the standard interpretation of states and the collapse dynamics.

2.3 *Von Neumann's formulation of quantum mechanics*

Von Neumann said that the object of his book *Mathematical Foundations of Quantum Mechanics* (1932) was to present a unified and mathematically rigorous formulation of the new quantum mechanics (von Neumann 1955, p. vii). While von Neumann took Dirac's formulation of quantum mechanics to be the best so far, he was critical of the mathematics. He complained that Dirac's formulation of quantum mechanics 'in no way satisfies the requirements of mathematical rigor—not even if these are reduced in a natural and proper fashion to the extent common elsewhere in theoretical physics' (p. ix). The problem was that in order to represent continuous physical quantities like position Dirac had introduced 'improper' functions to represent states where a system determinately had a particular value for the quantity. The probability distribution of a particle in an eigenstate of being at position a, for example, was represented by a function $\delta(a - x)$ that was supposed to be zero for all $x \neq a$ and yet integrate to one. Von Neumann noted that there were no such functions. He said that he would have no objection if such concepts could be shown to be intrinsically necessary for quantum mechanics, but he believed that quantum mechanics did not need bad mathematics.

In order to get a clear mathematical formulation of quantum mechanics, von Neumann needed some way of describing interactions between physical systems, including the interaction between a measuring device and its object system. Consequently, he provided one of the first clear statements of the quantum measurement problem and one of the first attempts to solve it—or, perhaps better, to argue that it was not a real problem.

Before discussing what von Neumann had to say about measurement, however, I shall consider his description of the standard collapse theory and how it should be interpreted.

2.3.1 *The standard theory*

The first thing that von Neumann does in his presentation of quantum mechanics is to describe the Hilbert-space formalism (see Appendix A for a summary of his description). He then discusses two general approaches for using the mathematical formalism to represent quantum-mechanical states and the dynamics.

On the Heisenberg picture the physical state is represented by a constant vector in an appropriate Hilbert space. Which space is appropriate depends on what physical quantities one is interested in representing. Observable properties are represented by Hermitian operators on the space that evolve over time. I shall describe the Heisenberg picture in more detail later (Chapter 8).

On the Schrödinger picture, the picture I shall use most of the time, the quantum-mechanical state of a system is represented by a unit-length vector in Hilbert space. Again, just as on the Heisenberg picture, the space one chooses will depend on what physical quantities one is interested in representing. But here, the state vector evolves according to the usual linear dynamics, which, if one is interested in position, is just the time-dependent Schrödinger equation (2.1). If $\hat{H}$ is time-independent, then the solution of this differential equation is

$$\phi(t) = e^{\frac{-2\pi i}{h}(t-t_0)\hat{H}} \psi(t_0), \qquad (2.5)$$

where the operator $e^{\frac{-2\pi i}{h}(t-t_0)\hat{H}}$ is unitary (von Neumann 1955: 208).[11] Since the vectors representing physical states are unit-length and since multiplying these by a constant has no physical meaning in the representation, one might picture the deterministic dynamics by imagining a point moving on a unit hemisphere in a continuous and perfectly deterministic way (a *hemisphere* because ψ represents the same physical state as $-\psi$).

Each physical quantity that one might observe corresponds to a Hermitian operator on an appropriate Hilbert space (von Neumann 1955: 200). If a system S is in the state ψ and if ψ is an eigenstate of the observable O with eigenvalue λ (that is, if $O\psi = \lambda\psi$), then if one measures O of S, then one is guaranteed to get the result λ. This tells us how to predict the result of a measurement whenever a system is in an eigenstate of the observable being measured. In general, however, a system will not be in an eigenstate of the observable being measured, so we need to know what happens when a measurement is made in these situations. In order to explain the role that measurement plays in the theory, von Neumann discussed his commitment to Born's rule, state completeness, and the fundamentally acausal nature of quantum mechanics.

Von Neumann was committed to *Born's rule* for the simple reason that it made the right statistical predictions. Born's rule says that the probability of a measurement of the physical observable O yielding the result λ when the system is initially in the state ψ is equal to $|\langle\phi|\psi\rangle|^2$, where ϕ is a unit-length eigenvector of O corresponding to the eigenvalue λ (von Neumann 1955: 216–17). Von Neumann was also committed to the principle of *state completeness*: everything that is in fact true of a physical system is determined by the element of Hilbert space that represents the system's state (196–7). And as Dirac had already argued, Born's rule together with the principle of state completeness requires one to accept the standard

---

[11] A unitary operator is a linear operator that will keep the sum of the quantum probabilities represented by the state equal to one. Here it has the effect of simply rotating the state vector.

eigenvalue–eigenstate link: a physical system has a particular physical property if and only if it is in an eigenstate of having that property.

It is this eigenvalue–eigenstate link that provides the standard interpretation of states. On the standard interpretation of states, a system only determinately has a property if it is in an eigenstate of having that property. It is not that it either has or does not have the property and that we just do not know which, but rather a system that fails to be in an eigenstate of having a particular property or an eigenstate of not having the property is in a *superposition* of having and not having the property, and so it does not determinately have the property, does not determinately not have the property, does not determinately both have and not have the property, and does not determinately neither have nor not have the property.

Since Born's rule makes the right statistical predictions, one would naturally want a formulation of quantum mechanics that accommodates this rule. But if one accepts Born's rule, then it seems that the only way to avoid the standard interpretation of states is to deny the principle of state completeness. So just how strongly was von Neumann committed to state completeness?

While von Neumann assumed for the purposes of his work that everything that can be said about the state of a physical system must be derived from the vector that represents its state, contrary to popular wisdom he did not believe that the standard quantum-mechanical state was necessarily complete. Rather, he thought that the principle of state completeness, the standard interpretation of states, and the acausal nature of quantum mechanics might ultimately prove to be provisional features of quantum mechanics. On the other hand, he also thought that sacrificing the principle of state completeness (and thus the standard interpretation of states) or demanding a deterministic theory would require a radical change in quantum mechanics, one that no one knew how to make.

Von Neumann seriously considered the possibility that the stochastic nature of quantum mechanics might be explained by appealing to a more fundamental or complete description of the world (as Einstein, for one, had argued). On this approach, one would deny state completeness and claim that

In reality, [the element of Hilbert space] ϕ does not determine the state exactly. In order to know this state absolutely, additional numerical data are necessary. That is, the system has other characteristics or coordinates in addition to ϕ. If we were to know all of these then we could give the values of all physical quantities exactly and with certainty. (von Neumann 1955: 209)

The hypothetical additional data that are to be added to the standard state description to make it complete are customarily called hidden parameters or hidden variables.

Whether or not an explanation of this type, by means of hidden parameters, is possible for quantum mechanics is a much discussed question. The view that it will sometime be answered in the affirmative has at present prominent representatives. If it were correct, it would brand the present form of the theory as provisional, since the description of states would then be incomplete. (209–10)

And he concluded that 'an introduction of hidden parameters is certainly not possible without a basic change in the present theory' (210), and, later in his book, he gave his famous argument against hidden-variable theories (313–25). But again, even here von Neumann rejected the hidden-variable approach to quantum mechanics not because he had shown that no such theory was possible, but rather because he had shown that it would require a very different sort of theory from the sort that physicists were currently considering.[12]

Much of the debate in the late 1920s concerning quantum mechanics was focused on whether or not it was possible to formulate an empirically adequate theory that satisfied the principle of sufficient cause (that was deterministic), but this debate was intimately connected to the question of how to interpret states. If one is committed to the principle of state completeness, then the probabilities that occur in quantum mechanics cannot be the result of our lack of information; rather, as von Neumann put it, these probabilities must result from the fact that nature itself has disregarded the principle of sufficient cause. If S_1 and S_2 are in the same physical state, then they must have the same physical properties, so there can be no reason (or cause) for one system behaving one way and the other system behaving another way, but quantum mechanics tells us (and our experience agrees) that such systems might in fact behave differently even if they are in exactly the same quantum-mechanical state (an x-spin

---

[12] Von Neumann allowed for the possibility that there was 'a more precise analysis of quantum mechanics' that would enable us to introduce hidden variables to explain the statistical properties of our observations, but for the time being, he said, 'we shall abandon this type of explanation' and 'admit as a fact that those natural laws that govern the elementary processes ... are of a statistical nature' (210). And he attributed this proposal to Born. 'This concept of quantum mechanics, which accepts its statistical nature in the actual form of the laws of nature, and which abandons the principle of causality, is the so called statistical interpretation. It is due to M. Born [1926] and is the only consistently enforceable interpretation of quantum mechanics today' (210).

measurement of a system in an eigenstate of z-spin can result in either x-spin up or x-spin down). Von Neumann noted that if one believed that two such systems might react differently when one measured some physical quantity, then one would typically have good reason to deny that they really were initially in the same state. He none the less argued that, while one might at first be sceptical, certain difficulties in quantum mechanics 'permit no other way out' but to deny that the laws of nature are causal (von Neumann 1955: 302–3): that is, we should take the quantum-mechanical state as complete and then deny that there is any physical cause for the difference in the behaviour of such systems.

But again, von Neumann's position was not as dogmatic as this sounds. While he argued that 'the only formal theory existing at the present time which orders and summarizes our experiences in this area [microscopic physics] in a half-way satisfactory manner, i.e. quantum mechanics, is in compelling logical contradiction with causality' (von Neumann 1955: 327), he also held that quantum mechanics was incomplete and perhaps even false:

Of course it would be an exaggeration to maintain that causality has hence been done away with: quantum mechanics has, in its present form, several serious lacunae, and it may even be that it is false, although this later possibility is highly unlikely, in the face of its startling capacity in the qualitative explanation of general problems, and in the quantitative calculation of special ones. In spite of the fact that quantum mechanics agrees well with experiment, and that it has opened up for us a qualitatively new side of the world, one can never say of the theory that it has been proved by experience, but only that it is the best known summarization of experience. However, mindful of such precautions, we may still say that there is at present no occasion and no reason to speak of causality in nature—because no experiment indicates its presence, since the macroscopic are unsuitable in principle, and the only known theory which is compatible with our experience relative to elementary processes, quantum mechanics, contradicts it. (327–8)

While causality is an 'age-old way of thinking of all mankind', it is not a logical necessity, and von Neumann did not believe that it was reasonable to sacrifice a physical theory that made such good empirical predictions for its sake (328). While he granted that quantum mechanics was a troubled theory, since there was nothing better to put in its place, and since any change that would make it more intuitive would require a very different theory, he considered his job to be to try to make sense of the theory that the physicists had developed so far.

Given his commitment to state completeness and the standard interpretation of states, von Neumann concluded that 'if the system is initially found in a state in which the values of R cannot be predicted with certainty, then this state is transformed by a measurement M of R . . . to another state: namely one in which the value of R is uniquely determined'. That is, sometime during a measurement, the state of the system being measured must instantaneously, randomly, and nonlinearly jump to an eigenstate of the observable being measured. And the Born rule requires the probability that a system initially in the state ψ will end up in the eigenstate ϕ_n to be $|\langle \psi | \phi_n \rangle|^2$. This, finally, was von Neumann's description of what happens during a measurement.[13] He concluded then that 'We therefore have two fundamentally different types of interventions which can occur to a system S . . . First the arbitrary changes by measurements', where the state ψ evolves discontinuously, randomly, and nonlinearly to an eigenstate of the observable being measured ϕ_n with probability $|\langle \psi | \phi_n \rangle|^2$. 'Second, the automatic changes which occur with the passage of time', where the state $\psi(t_0)$ evolves continuously and deterministically as described by the time-dependent Schrödinger equation to the state $U(t - t_0)\psi(t_0)$ at time t. Von Neumann called these Process 1 and Process 2, respectively.[14]

Having two dynamical laws, however, poses a problem for the theory, which von Neumann himself immediately recognized. Since the linear Schrödinger equation (Process 2) allows one to describe the interaction between two systems if one knows their initial states and the Hamiltonian $\hat{H}$, it alone should be sufficient to describe the interaction between a measuring device and the system being measured. Indeed, what else could a measurement be other than an ordinary interaction between two physical systems? As von Neumann confessed, 'one should expect that [Process 2] would suffice to describe the intervention caused by a measurement: Indeed, a physical intervention can be nothing else than the temporary insertion of a certain energy coupling into the observed system, i.e., the introduction of an appropriate time dependency of $\hat{H}$ (prescribed by the observer)' (1955: 352). And he concluded that

We have then answered the question as to what happens in the measurement of a quantity R, under the above assumptions for its operator R. To be sure the 'how' remains unexplained for the present. This discontinuous transition from ψ into one of the states $\phi_1, \phi_2, \ldots \ldots$ is certainly not of the type described by the time

---

[13] Like Dirac, von Neumann took the repeatability of measurements to have provided the best empirical evidence for the random collapse of the state on measurement.

[14] The $U(t - t_0)$ is a family of linear (unitary), time-evolution operators that represent how the quantum-mechanical state evolves under the time-dependent Schrödinger equation.

dependent Schrödinger equation. The latter always results in a continuous change of ψ, in which the final result is uniquely determined and dependent on ψ. (217)

2.3.2 *A summary of the theory*

Setting aside the question of precisely how and when collapses occur, we now have most of quantum mechanics on the Schrödinger picture. Together with a rule for how to represent the states of composite systems and independent properties, the principles of the theory are as follows:

1. *Representation of states.* The state of a physical system S is represented by an element $|\psi\rangle_S$ of unit length in a Hilbert space $\mathcal{H}$.

2. *Representation of observables.* Every physical observable O is represented by a Hermitian operator $\hat{O}$ on the Hilbert space, and every Hermitian operator on the Hilbert space corresponds to some complete observable.

3. *Interpretation of states.* A system S has a determinate value for observable O if and only if it is in an eigenstate of O: that is, S has a determinate value for O if and only if $\hat{O}|\psi\rangle_S = \lambda|\psi\rangle_S$, where $\hat{O}$ is the Hermitian operator corresponding to O, $|\psi\rangle_S$ is the vector representing the state of S, and the eigenvalue λ is a real number. In this case, one would with certainty get the result λ if one measured O of S.

4. *Laws of motion*

I. Linear dynamics. If no measurement is made of a physical system, it will evolve in a deterministic, linear way: if the state of S is given by $|\psi(t_0)\rangle_S$ at time t_0, then its state at a time t_1 will be given by $\hat{U}(t_0, t_1)|\psi(t_0)\rangle_S$, where $\hat{U}(t_0, t_1)$ is a unitary operator on $\mathcal{H}$ that depends on the energy properties of S.

II. Nonlinear collapse dynamics. If a measurement is made of the system S, it will instantaneously and nonlinearly jump to an eigenstate of the observable being measured (a state where the system has a determinate value of the property being measured). If the initial state is given by $|\psi\rangle_S$ and $|\chi\rangle_S$ is an eigenstate of O, then the probability of S collapsing to $|\chi\rangle_S$ is equal to $|\langle\psi|\chi\rangle|^2$ (the magnitude of the projection of the pre-measurement state onto the eigenstate). That is, if a measurement is made, then the system instantaneously and randomly jumps from the initial superposition to an eigenstate of the observable being measured:

$$\psi = \sum_k c_k|\psi_k\rangle \longrightarrow |\psi_j\rangle, \qquad (2.6)$$

where $|c_j|^2$ is the probability of ending up in the eigenstate $|\psi_j\rangle$.

5. *Composition rule*. If system S_1 is represented by an element $|\phi\rangle$ of $\mathcal{H}_1$ and S_2 by an element $|\psi\rangle$ of $\mathcal{H}_2$, then the composite system $S_1 + S_2$ is represented by an element $|\phi\rangle \otimes |\psi\rangle$ (which I will usually write as just $|\phi\rangle|\psi\rangle$) of $\mathcal{H}_1 \otimes \mathcal{H}_2$ (the direct-product space). Similarly, if property P_1 of system S is represented by an element $|\phi\rangle$ of $\mathcal{H}_1$ and an independent (quantum-mechanically compatible) property P_2 of S by an element $|\psi\rangle$ of $\mathcal{H}_2$, then both properties can be simultaneously represented by an element $|\phi\rangle \otimes |\psi\rangle$ of $\mathcal{H}_1 \otimes \mathcal{H}_2$. A collapse of the state of a composite system selects one term in the superposition of product states that contain the eigenstates of the observable being measured (and the examples below should help to make it clear what this means).

2.4   How the theory works

The standard theory makes very accurate empirical predictions and goes a considerable way in explaining the quantum phenomena we see in the laboratory. In order to see how the theory works, we will consider the spin properties of spin-$\frac{1}{2}$ particles again.

The Hilbert space one uses to represent the state of a system is typically complex-valued and can be infinite-dimensional, but if one is interested in only its x-spin and z-spin properties, then the spin state of an electron can be represented by a unit-length vector in ordinary 2-dimensional Euclidean space (if one were also interested in the y-spin of the electron, then one would have to represent its state in C^2).

An electron, like any other system, determinately has a property if and only if it is in an eigenstate of having that property. There are two x-spin eigenstates $|\uparrow_x\rangle$ and $|\downarrow_x\rangle$ (corresponding to the electron determinately being x-spin up and determinately being x-spin down) and two z-spin eigenstates $|\uparrow_z\rangle$ and $|\downarrow_z\rangle$ (corresponding to the electron determinately being z-spin up and determinately being z-spin down). One can account for the experimental results if one represents these eigenstates by the following vectors:

$$|\uparrow_x\rangle = \begin{pmatrix} 1 \\ 0 \end{pmatrix}, |\downarrow_x\rangle = \begin{pmatrix} 0 \\ 1 \end{pmatrix}, |\downarrow_z\rangle = \begin{pmatrix} 1/\sqrt{2} \\ 1/\sqrt{2} \end{pmatrix}, |\downarrow_z\rangle = \begin{pmatrix} 1/\sqrt{2} \\ -1/\sqrt{2} \end{pmatrix}. \quad (2.7)$$

It follows that the x-spin eigenstates are superpositions of z-spin eigenstates and that z-spin eigenstates are superpositions of x-spin eigenstates:

$$|\uparrow_x\rangle = 1/\sqrt{2}(|\uparrow_z\rangle + |\downarrow_z\rangle) \quad (2.8)$$

$$|\downarrow_x\rangle = 1/\sqrt{2}(|\uparrow_z\rangle - |\downarrow_z\rangle) \tag{2.9}$$

$$|\uparrow_z\rangle = 1/\sqrt{2}(|\uparrow_x\rangle + |\downarrow_x\rangle) \tag{2.10}$$

$$|\downarrow_z\rangle = 1/\sqrt{2}(|\uparrow_x\rangle - |\downarrow_x\rangle). \tag{2.11}$$

Given the choice of how to represent the spin eigenstates, one might represent the x-spin operator as

$$\sigma_x = \begin{pmatrix} 1 & 0 \\ 0 & -1 \end{pmatrix}. \tag{2.12}$$

If $\sigma_x|\psi\rangle_S = \lambda|\psi\rangle_S$, then if $\lambda = 1$, then P is determinately x-spin up, and if $\lambda = -1$, then P is determinately x-spin down.

One might similarly represent the z-spin operator as

$$\sigma_z = \begin{pmatrix} 0 & 1 \\ 1 & 0 \end{pmatrix}. \tag{2.13}$$

And if $\sigma_z|\psi\rangle_S = \lambda|\psi\rangle_S$, then if $\lambda = 1$, then P is determinately z-spin up, and if $\lambda = -1$, then P is determinately z-spin down.

Now, suppose we measure the x-spin of an electron P initially in the state $|\uparrow_z\rangle$. Since it is not in an eigenstate of x-spin, according to the standard interpretation of states, P does not even have a determinate x-spin to be measured. In order to get a determinate x-spin, P instantaneously, randomly, and nonlinearly jumps into an eigenstate of x-spin. The collapse dynamics tells us that the probability of jumping to $|\uparrow_x\rangle$ is $|\langle\uparrow_z|\uparrow_x\rangle|^2 = 1/2$ and the probability of jumping to $|\downarrow_x\rangle$ is $|\langle\uparrow_z|\downarrow_x\rangle|^2 = 1/2$. This explains why about half of the electrons that start in a z-spin up eigenstate are found to be x-spin up and about half are found to be x-spin down. It also explains why one will get the same x-spin result if a second x-spin measurement is made: while the electron does not begin with a determinate x-spin, the first measurement would give it a determinate x-spin, and it would have this x-spin until it is disturbed or until a measurement incompatible with x-spin (like a z-spin measurement) is made.[15]

Suppose we get x-spin down as the measurement result, then decide to measure the z-spin of the electron. Since the electron is no longer in an

---

[15] I am assuming throughout that spin properties are a constant of motion. Whether or not this is the case in a particular physical situation depends on the Hamiltonian that describes the interaction between the spin-$\frac{1}{2}$ particle and its environment. There are, of course, Hamiltonians that would cause the spin state of the particle to evolve in interesting ways.

eigenstate state of z-spin, the collapse dynamics tells us that the electron will randomly jump into an eigenstate of z-spin and that the probability of jumping to $|\uparrow_z\rangle$ is $|\langle\downarrow_x|\uparrow_z\rangle|^2 = 1/2$ and the probability of jumping to $|\downarrow_z\rangle$ is $|\langle\downarrow_x|\downarrow_z\rangle|^2 = 1/2$.

Note that the x-spin measurement causes the electron to jump to an eigenstate of x-spin, which is a state where it has no determinate z-spin, and a subsequent z-spin measurement causes the electron to jump back to a state where it has a determinate z-spin, but since these jumps are random, it may not end up in the z-spin eigenstate in which it started. The collapse dynamics then explains how a z-spin measurement makes the result of a subsequent x-spin measurement entirely uncertain regardless of how careful one is not to disturb the system.

In the two-path interference experiment described earlier (in Section 1.2), the initial state of the z-spin up electron at location S might be written as

$$|\uparrow_z\rangle = 1/\sqrt{2}(|\uparrow_x\rangle_P + |\downarrow_x\rangle_P)|S\rangle_P. \tag{2.14}$$

If the x-spin sorter would send an x-spin up electron along path A and an x-spin down electron along path B, then it would send a z-spin up along a superposition of the two paths. More specifically, after passing through the x-spin sorter, the state of the electron will be a superposition of it being x-spin up and on path A and it being x-spin down and on path B, and one might write this state as

$$1/\sqrt{2}(|\uparrow_x\rangle_P|A\rangle_P + |\downarrow_x\rangle_P|B\rangle_P). \tag{2.15}$$

This is a state where the electron's spin properties and position properties are entangled and where it consequently fails to have any determinate pure spin or pure position properties (according to the standard interpretation of states). If one does not look for the electron on either path, then the linear dynamics will continue to describe the time-evolution of its state. If the state of the electron were $|\uparrow_x\rangle_P|A\rangle_P$, then it would evolve to $|\uparrow_x\rangle_P|I\rangle_P$; and if its state were $|\downarrow_x\rangle_S|B\rangle_P$, then it would evolve to $|\downarrow_x\rangle_P|I\rangle_P$ (in the same time since the experiment is set up so that the two paths are the same length). It follows, then, from the linearity of the dynamics that the above superposition would evolve to

$$1/\sqrt{2}(|\uparrow_x\rangle_P|I\rangle_P + |\downarrow_x\rangle_P|I\rangle_P). \tag{2.16}$$

The position of P is no longer entangled with its spin, which we can represent by writing this state as

$$1/\sqrt{2}(|\uparrow_x\rangle_P + |\downarrow_x\rangle_P)|I\rangle_P, \tag{2.17}$$

which is just another way of saying that the electron is now determinately at position I and again is determinately z-spin up (and we have explained the interference statistics!).

If, however, we look for the electron while it is in the entangled super-position above, then the collapse dynamics kicks in, and it will jump to an eigenstate of position, which will also put it in the corresponding eigen-state of x-spin. If we look at path B, for example, then either it will or will not be found on that path. The probability of the state collapsing to $|\uparrow_x\rangle_P|A\rangle_P$ (in which case we would not see it on path B) is 1/2 and the probability of the state collapsing to $|\downarrow_x\rangle_P|B\rangle_P$ (in which case we would see it on path B) is also 1/2. But, in either case, the result of a subsequent z-spin measurement would be completely random. The fact that observing path B causes the electron to collapse to an eigenstate of position *even if it is not found on path B* explains why looking for the electron can affect its behaviour even when one fails to find the electron (and we have explained the A-or-B statistics!).

It is important to note the role played by the collapse postulate here: (1) it is the collapse of the state onto an eigenstate of the observable that we are measuring that explains why we get a determinate result to the measurement, (2) it is also what explains why we get the same result if we repeat the measurement without disturbing the system, and (3) the collapse dynamics is essential to the standard explanation of the statistical properties of our experimental results. And remember: Everett wants to drop the collapse dynamics from the standard theory and deduce exactly the same results as subjective appearances!

2.5   Two discontents

In spite of its remarkable empirical successes, Einstein continued to argue against quantum mechanics. He did not like the claim that the quantum-mechanical state was complete, nor did he like the special acausal role that measurement was supposed to play in the theory. He took the col-lapse dynamics to be an artefact of an incorrect understanding of the quantum-mechanical state, and he was worried about its compatibility with relativity.

In 1935 Albert Einstein, Boris Podolsky, and Nathan Rosen (EPR) argued that the quantum-mechanical description of physical reality could not be considered to be complete. While the worry was essentially the same as that expressed by Einstein at the 1927 Solvay Congress, the argu-ment was more sophisticated and careful. The basic argument was that

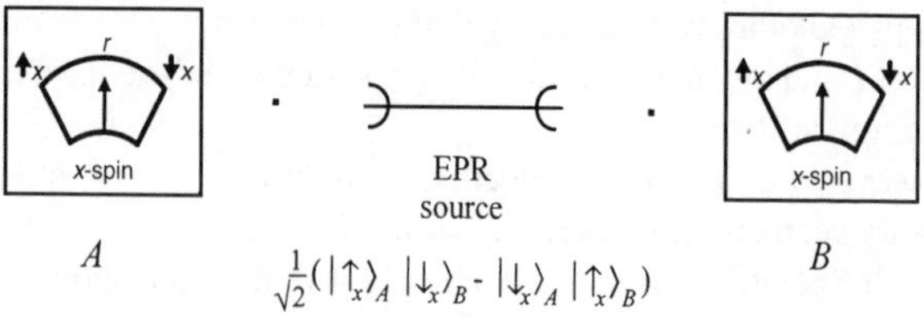

$$\frac{1}{\sqrt{2}}(|\uparrow_x\rangle_A |\downarrow_x\rangle_B - |\downarrow_x\rangle_A |\uparrow_x\rangle_B)$$

FIG. 2.3 EPR setup.

since one can without in any way disturbing the state of a system (assuming that all disturbances are local) predict with certainty the value of a physical quantity that is not determined by the (pre-measurement) quantum-mechanical state, the quantum-mechanical state must be an incomplete description of the physical properties of the system.

Consider two particles, one at location A and one at location B, in an entangled spin state $1/\sqrt{2}(|\uparrow_x\rangle_A |\downarrow_x\rangle_B - |\downarrow_x\rangle_A |\uparrow_x\rangle_B)$ (Fig. 2.3). If one measures the x-spin of the particle at A, then one can predict with certainty the result of an immediately subsequent x-spin measurement of the particle at B (it will be exactly the opposite of the result at A). But if the value of a physical quantity can be predicted with certainty without the system being disturbed, then EPR held that there must exist an element of physical reality corresponding to that physical quantity. It is the existence of this element of reality that explains how we are able to predict the result of the second measurement with certainty. If one assumes that all physical interactions are local (something that Einstein seems to have associated with the correctness of relativity), then one's measurement at A could in no way have disturbed the state of the particle at B, so there must have been an objectively real, determinate value for the x-spin of B all along, even before our measurement of A. But the x-spin of B is not determined by the initial quantum-mechanical state description of the two-particle system—indeed, the entangled spin state is perfectly symmetric with respect to both possible x-spin results at B. Consequently, there must exist elements of physical reality that are not specified by the standard quantum-mechanical description, which means that it provides only an incomplete description of the real physical world (or quantum mechanics is nonlocal and thus, on Einstein's view, somehow incompatible with relativity).[16]

---

[16] The original EPR paper is reprinted in Wheeler and Zurek (1983: 138–41).

Schrödinger was also unhappy with the standard formulation of quantum mechanics. In light of Born's statistical interpretation of the wave function and the associated collapse dynamics, he even came to regret his role in the development of the theory: after debating the acausal nature of quantum mechanics with Bohr, Schrödinger exclaimed: 'If one has to stick to this damned quantum jumping, then I regret having ever been involved in this thing' (Jammer 1974: 344, ref. 4; Pais 1988: 261). The 1935 EPR paper motivated Schrödinger to write a response describing the contemporary situation in quantum mechanics. He presented his paper as a confession.[17]

According to the standard theory of quantum mechanics, Schrödinger explained, the quantum-mechanical state describes the type and degree of the 'blurring' of physical properties. While one might not worry about this indeterminacy of physical properties 'so long as the blurring is confined to atomic scale', where we cannot observe exactly what is happening anyway, Schrödinger proceeded to show why quantum *microscopic* indeterminacy must end up generating *macroscopic* indeterminacy via the linear dynamics (Wheeler and Zurek 1983: 156). This is the point of his famous cat experiment:

> One can even set up quite ridiculous cases. A cat is penned up in a steel box, along with the following diabolical device (which must be secured against direct interference by the cat): in a Geiger counter there is a tiny bit of radioactive substance, *so* small, that *perhaps* in the course of one hour one of the atoms decays, but also, with equal probability, perhaps none; if it happens, the counter tube discharges and through a relay releases a hammer which shatters a small flask of hydrocyanic acid. If one has left this entire system to itself for an hour, one would say that the cat still lives *if* meanwhile no atom has decayed. The first atomic decay would have poisoned it. The ψ-function would express this by having in it the living and the dead cat (pardon the expression) mixed or smeared out in equal parts. (157)

It is not the blurred model of reality itself that led Schrödinger to conclude that quantum mechanics cannot be taken as an accurate description of the physical world. He believed that there was nothing inherently unclear or contradictory about such a model, especially if it is applied to microscopic events alone. But since microscopic indeterminacy will invariably infect macroscopic properties and since macroscopic properties always appear to us to be perfectly determinate, quantum mechanics

---

[17] Schrödinger's 1935 paper is translated and reprinted in Wheeler and Zurek (1983: 152–67).

must somehow *resolve the macroscopic indeterminacy in order to account for our determinate experience*, and it was the resolution of macroscopic indeterminacy by way of the collapse dynamics that Schrödinger did not like. He argued that the abrupt change in the quantum-mechanical state ψ by measurement implies that ψ cannot be taken to represent the real physical state of affairs (158). Part of the argument was that if measurements are perfectly ordinary physical interactions like any other, then the time-evolution of the complete and accurate state description would have to be described by the same dynamics whether a measurement was being made or not. But since the standard theory requires a different dynamical law to describe the time-evolution of the state ψ when a measurement is being made, ψ cannot be the complete and accurate state description.

> Now it was previously stated ... and explained ... that any *measurement* suspends the law that otherwise governs the continuous time-dependence of the ψ-function and brings about in it a quite different change, not governed by any laws, but rather dictated by the result of the measurement. But laws of nature differing from the usual ones cannot apply during a measurement, for objectively viewed it is a natural process like any other, and it cannot interrupt the orderly course of natural events. Since it does interrupt that of the ψ-function the latter ... can*not* serve ... as an experimentally verifiable representation of an objective reality. (160)

Note that this is somewhat different from von Neumann's worry. Here the argument is that since ψ must follow an *unnatural* evolution during measurement (in order for us to account for our determinate experience) it cannot be taken as a complete representation of the *natural* physical state.

Schrödinger also shared Einstein's worry about the compatibility of quantum mechanics and relativity. He argued that 'the conceptual joining of two or more systems into *one* encounters great difficulty as soon as one attempts to introduce the principle of special relativity into Q. M.' In any case, he thought that anyone who reflected 'on the apparent jumping around of the ψ-function' and the 'antinomies of entanglement' ought to worry about the compatibility of quantum mechanics with relativity (Wheeler and Zurek 1983: 167).

And Schrödinger and Einstein were right to worry. The standard collapse theory, at least, really is incompatible with the theory of relativity in a perfectly straightforward way: the collapse dynamics is not Lorentz-covariant. When one finds an electron, for example, its wave function instantaneously goes to zero everywhere except where one found it. If this did not happen, then there would be a nonzero probability of finding the electron in two places at the same time in the measurement frame. The problem is that we cannot describe this process of the wave function going to zero almost everywhere *simultaneously* in a way that is compatible with relativity. In relativity there is a different standard of simultaneity for each inertial frame, but if one chooses a particular inertial frame in order to describe the collapse of the wave function, then one violates the requirement that all physical processes must be described in a frame-independent way.[18] There have been several attempts to express the collapse dynamics in a frame-independent, or more specifically a Lorentz-covariant, way, but these attempts have encountered their own problems.[19]

Given the standard theory's problems, Schrödinger concluded that 'the simple procedure provided ... by the non-relativistic theory is perhaps after all only a convenient calculational trick, but one that today, as we have seen, has attained influence of unprecedented scope over our most basic attitude toward nature' (Wheeler and Zurek 1983: 167). Schrödinger was proud of the empirical successes of the theory, but he thought that there was something seriously wrong with the way that the standard theory was supposed to account for our determinate experience.

Einstein later described his own worries about quantum mechanics as being akin to Schrödinger's. In a letter to Schrödinger dated 22 December 1950 Einstein wrote:

You are the only contemporary physicist, besides Laue, who sees that one cannot get around the assumption of reality—if only one is honest. Most of them simply do not see what sort of risky game they are playing with reality—reality is something independent of what is experimentally established. They somehow believe that quantum theory provides a description of reality, and even a *complete* description; this interpretation is, however, refuted, most elegantly by your system of radioactive atom + Geiger counter + amplifier + charge of gun powder + cat in a box, in which the ψ-function of the system contains the cat both alive and blown to bits. Is the state of the cat to be created only when the physicist investigates the situation at some definite time? Nobody really doubts that the presence or absence of the cat is something independent of the act of observation. But then the description by means of the ψ-function is certainly incomplete, and there must be a more complete description. (Przibram 1967: 39)

---

[18] One might be tempted to argue here that the evolution of the quantum-mechanical state is not really a physical process. But if the quantum-mechanical state is not the *physical state*, then what is it? It is worth remembering that the ignorance interpretation doesn't work.

[19] See Aharonov and Albert (1981). See also Fleming (1996). On Gordon Fleming's proposal there are no simple matters of fact about the physical state in a space-time region; rather, the physical state in a space-time region is *hyperplane-dependent*.

Einstein also repeatedly tried to explain his worries about quantum mechanics to his friend Max Born. Born believed that Einstein's rejection of the theory was primarily due to his commitment to determinism. For his part, Einstein seems to have become rather frustrated with Born's failure to see how serious the conceptual problems faced by the theory really were.

In 1954 Pauli decided that he would step in and explain Einstein's position to Born. Pauli wrote that Einstein's true worry was not the lack of determinism but rather, just as Einstein had explained to Schrödinger, that quantum mechanics failed to be realistic—that the quantum-mechanical state could not be taken as a *complete* description of the actual physical state of affairs (Pauli 1971: 221). If the usual linear dynamics is right, then even macroscopic objects can be in superpositions of having very different positions, so it cannot be on the basis of their quantum-mechanical states that they have determinate positions (221–2). Further, Pauli argued, 'it is not reasonable to invent a causal mechanism according to which "looking" fixes the position' (222). Hence the quantum-mechanical state cannot be a complete and accurate objective description of reality (223).

Pauli thought that many of Einstein's worries, and this one in particular, were like the question concerning how many angels might sit on the head of a pin. His own position here, however, is interesting. Pauli seems to have agreed with Einstein that it is unreasonable to invent a mechanism for fixing the position of a particle when we look at it. But if this is right, then Pauli must have also found the collapse dynamics unsatisfactory. But then how did he understand quantum mechanics?

Pauli argued that it was too much to ask for natural laws to explain why we experience what we do. The reason was that appearances, while not within the control of an observer, are created outside nature and consequently not within the domain of natural law:

The appearance of a definite position x_0 during a subsequent observation (for example 'illumination of the place with a shaded lantern') ... and the statement 'the particle is there', is then regarded as being a 'creation' existing outside the laws of nature, even though it cannot be influenced by the observer. The natural laws only say something about the *statistics* of these acts of observation. (Pauli 1954: 223)

Rather than promoting the standard collapse formulation of quantum mechanics, then, Pauli seems to have had in mind a version of quantum mechanics where the usual linear dynamics always correctly describes the time-evolution of the state of every physical system and where our determinate experiences are explained by *extra-physical* matters of fact.

To me, this sounds very much like the sorts of theory we will discuss in Chapter 7 (the single-mind, many-minds, and correlations-without-correlata theories in particular), where determinate appearances are to be explained by mental (or some other extra-physical facts in the case of the correlations-without-correlata view) rather than by physical facts.

2.6   *Von Neumann's psychophysical parallelism and Wigner's friend*

Contrary to Pauli's (much later) intuitions, it was important to von Neumann that a satisfactory physical theory be able to provide some sort of account of subjective perception:

it is a fundamental requirement of the scientific viewpoint—the so-called principle of the psychophysical parallelism—that it must be possible so to describe the extraphysical process of the subjective perception as if it were in reality in the physical world—i.e., to assign to its parts equivalent physical processes in the objective environment, in ordinary space. (von Neumann 1955: 418–19)

That is, a good physical theory should allow for the possibility of describing a determinate sequence of physical events that might be identified with the sequence of mental events involved in performing an observation. Of course, it is not the job of a physical theory to give a fine-grained account of mental processes, but it should at least be compatible with the possibility of identifying a sequence of physical events with the sequence of mental events that constitutes a specific observation.

Von Neumann gave the example of measuring a temperature, and he traced the causal chain of events from the physical system whose temperature is being measured to the glass of the thermometer containing the mercury, to the length of the column of mercury, to the path of the light reflected off the column, to the image of the mercury column on the observer's retina, to the optical nerve tract, and finally to the chemical changes in the brain of the observer making the measurement (1955: 419). At some point in this chain, he concluded, we must say: 'and this is perceived by the observer'. Hence, 'we must always divide the world into two parts, the one being the observed system, the other the observer'. But, he argued, it is a consequence of the example of measuring temperature that 'the boundary between the two parts is arbitrary to a very large extent' (420). One could include the thermometer as a part of the observer, or in the other direction, one could exclude the observer's eyes. But if we are going to talk about the empirical predictions of a theory, then we must introduce the subjective observer and draw a boundary between the observer and the observed since 'experience only makes statements of this type: an observer has made a certain (subjective) observation; and never any like this: a physical quantity has a certain value' (420).

The point of course is that the analysis of a temperature measurement is supposed to make us worry less about the two dynamical laws required by quantum mechanics.

Now quantum mechanics describes the events which occur in the observed parts of the world, so long as they do not interact with the observing portion, with the aid of process 2., but as soon as such an interaction occurs, i.e., a measurement, it requires the application of process 1. The dual form is thereby justified. However, the danger lies in the fact that the principle of the psycho-physical parallelism is violated, so long as it is not shown that the boundary between the observed system and the observer can be displaced arbitrarily in the sense given above [in the temperature example]. (von Neumann 1955: 420–1)

That is, von Neumann wanted to show that it makes no empirical difference to the observer where Process 1 occurs in the chain of events leading to the observer perceiving a particular measurement result. He felt that this would imply that the boundary between the observer and the observed system was arbitrary, which was then supposed to show that quantum mechanics was compatible with the principle of psychophysical parallelism, that it allowed one to identify a sequence of physical events with the process of making an observation.

Von Neumann considered a measurement consisting of three physical systems: the system being observed P, the measuring device M, and the observer O. In this context, he wanted to show that the standard formulation of quantum mechanics makes the same empirical predictions if Process 1 applies to P and Process 2 applies to the interaction between P and $M + O$ as it does if Process 2 applies to $P + M$ and Process 1 applies to the interaction between $P + M$ and O.

Suppose then that observer O uses measuring device M to measure the x-spin of a spin-$\frac{1}{2}$ particle P. There are two possible results, *up* (corresponding to the state $|\uparrow\rangle_P$) and *down* (corresponding to the state $|\downarrow\rangle_P$). In order to specify the nature of the interaction between the systems it suffices here to specify two dispositions for M and two for O. I shall suppose here that the measuring device and the observer are perfect and that we know their initial states.[20]

---

[20] Von Neumann considers and rejects the claim that the quantum-mechanical statistics are somehow the result of incomplete knowledge of the quantum-mechanical state of the observer because it is a complex system. Since he does not find such an argument convincing,

If M is an x-spin measuring device, then its pointer will become correlated with the x-spin of the system it measures. I shall also suppose (without any loss in generality here) that in correlating its pointer M does not disturb the state of its object system. These two conditions entail precisely the same two dispositions described earlier:

(1) $|\text{ready}\rangle_M|\uparrow\rangle_P \longrightarrow |\uparrow\rangle_M|\uparrow\rangle_P$

and

(2) $|\text{ready}\rangle_M|\downarrow\rangle_P \longrightarrow |\downarrow\rangle_M|\downarrow\rangle_P$,

where $|\text{ready}\rangle_M$ represents a state where M is ready to make an x-spin measurement on P and $|\uparrow\rangle_M$ and $|\downarrow\rangle_M$ represent M in a state where its pointer reports the results x-spin up and x-spin down, respectively. Similarly, I shall suppose that the observer O correlates his brain state with the report made by the measuring device, and that this process does not disturb M or P:

(1) $|\text{ready}\rangle_O|\uparrow\rangle_M|\uparrow\rangle_P \longrightarrow |\uparrow\rangle_O|\uparrow\rangle_M|\uparrow\rangle_P$

and

(2) $|\text{ready}\rangle_O|\downarrow\rangle_M|\downarrow\rangle_P \longrightarrow |\downarrow\rangle_O|\downarrow\rangle_M|\downarrow\rangle_P$,

where $|\text{ready}\rangle_O$ represents a state where the observer is ready to look at M's pointer and read the result of the x-spin measurement and $|\uparrow\rangle_O$ and $|\downarrow\rangle_O$ represent O in a state where the observer believes that the result was x-spin up and believes that the result was x-spin down, respectively.

Now suppose that P begins in a superposition of the two x-spin eigenstates $1/\sqrt{2}(|\uparrow\rangle_P + |\downarrow\rangle_P)$ and M and O begin in their ready states. The initial state of the composite system $P + M + O$, then, is

$$|\text{ready}\rangle_O|\text{ready}\rangle_M\frac{1}{\sqrt{2}}(|\uparrow\rangle_P + |\downarrow\rangle_P). \qquad (2.18)$$

Now consider von Neumann's two cases. In the first case, we suppose that the collapse (Process 1) occurs when M interacts with P and that the interaction between O and M is given by the usual linear dynamics (Process 2) (P is the observed system and $O + M$ the observer). When M and P interact two things happen. First, P collapses to an eigenstate of the observable being measured with probabilities determined by Born's rule.

he stipulates that 'we will assume in all that follows that [the state of the observer] is completely known' (1955: 437–9).

In this case, there is a probability of $1/2$ that it will end up in an x-spin up eigenstate and a probability of $1/2$ that it will end up in an x-spin down eigenstate. So the composite system just before M's pointer becomes correlated with the x-spin of P will be represented by the statistical mixture:

$$p(|\text{ready}\rangle_O |\text{ready}\rangle_M |\uparrow\rangle_P) = 1/2$$
$$p(|\text{ready}\rangle_O |\text{ready}\rangle_M |\downarrow\rangle_P) = 1/2,$$

where the probabilities here are the *epistemic probabilities* associated with each state. The composite system really is in one or the other of the state in a (proper) statistical mixture like that above—we just do not know which state it is in. And after M's pointer becomes correlated with the x-spin of P, it will be represented by the mixture:

$$p(|\text{ready}\rangle_O |\uparrow\rangle_M |\uparrow\rangle_P) = 1/2$$
$$p(|\text{ready}\rangle_O |\downarrow\rangle_M |\downarrow\rangle_P) = 1/2.$$

And finally, after O's brain state becomes correlated with the position of M's pointer, it will be represented by the mixture:

$$p(|\uparrow\rangle_O |\uparrow\rangle_M |\uparrow\rangle_P) = 1/2$$
$$p(|\downarrow\rangle_O |\downarrow\rangle_M |\downarrow\rangle_P) = 1/2.$$

That is, there is a probability of $1/2$ that the final state of the composite system will be $|\uparrow\rangle_O |\uparrow\rangle_M |\uparrow\rangle_P$ and a probability of $1/2$ that it will be $|\downarrow\rangle_O |\downarrow\rangle_M |\downarrow\rangle_P$.

In the second case we suppose that the interaction between P and M is described by the usual linear dynamics (Process 2) and that the collapse occurs when $P + M$ interacts with O ($P + M$ is the observed system and only O the observer). In this case, it follows from the two dispositions of M and the linearity of the dynamics that the initial state (2.18) evolves to

$$|\text{ready}\rangle_O \frac{1}{\sqrt{2}} (|\uparrow\rangle_M |\uparrow\rangle_P + |\downarrow\rangle_M |\downarrow\rangle_P) \qquad (2.19)$$

as M's pointer becomes correlated with the x-spin of P. Then when O interacts with $P + M$ two things happen. First, the composite system randomly evolves to a state where M's pointer has a determinate position

$$p(|\text{ready}\rangle_O |\uparrow\rangle_M |\uparrow\rangle_P) = 1/2$$
$$p(|\text{ready}\rangle_O |\downarrow\rangle_M |\downarrow\rangle_P) = 1/2.$$

Then O's brain state becomes correlated with the position of the pointer

$$p(|\uparrow\rangle_O |\uparrow\rangle_M |\uparrow\rangle_P) = 1/2$$
$$p(|\downarrow\rangle_O |\downarrow\rangle_M |\downarrow\rangle_P) = 1/2.$$

So, just as in the first case, there is a probability of $1/2$ that the final state of the composite system will be $|\uparrow\rangle_O |\uparrow\rangle_M |\uparrow\rangle_P$ and a probability of $1/2$ that it will be $|\downarrow\rangle_O |\downarrow\rangle_M |\downarrow\rangle_P$. Von Neumann concludes, then, that it is empirically irrelevant where in the chain of events leading to a conscious measurement result the collapse Process 1 occurs. Thus, quantum mechanics poses no problem for the principle of psycho-physical parallelism.

While von Neumann showed that the predictions that the theory makes for the result of O's x-spin measurement do not depend on when Process 1 occurs in the (von Neumann) chain of measurement interactions, he did not show that *when it is applied* is empirically irrelevant in general. Indeed, it *does* generally make an empirical difference when a collapse occurs, which means that, rather than being a virtue, the fact that quantum mechanics does not tell us when collapses occur makes the theory incomplete *in an empirically significant way.*

Consider the physical state of $M + P$ in each case after M has interacted with P and just before O looks at the pointer. In the first case, the state is either $|\uparrow\rangle_M |\uparrow\rangle_P$ or $|\downarrow\rangle_M |\downarrow\rangle_P$. In the second case the state is $1/\sqrt{2}(|\uparrow\rangle_M |\uparrow\rangle_P + |\downarrow\rangle_M |\downarrow\rangle_P)$. And there are at least in principle experiments that can distinguish between either of the states in the first case and the state in the second case. Let A be an observable that has $1/\sqrt{2}(|\uparrow\rangle_M |\uparrow\rangle_P + |\downarrow\rangle_M |\downarrow\rangle_P)$ as an eigenstate with eigenvalue $+1$ and every orthogonal state as an eigenstate with eigenvalue -1 (given the correspondence that von Neumann postulates between physical properties and Hermitian operators, there must be just such an observable). If one were to measure this A observable of the composite system $M + P$, then one would get the result $+1$ about half of the time if the composite system was in either of the first two states, but one would get the result $+1$ with certainty if the composite system was in the second state. Consequently, *when* Process 1 occurs in the chain of measurement interactions makes an empirical difference, not for the results of x-spin measurements but for the results of A-type measurements.

Eugene Wigner recognized that when collapses occur makes an empirical difference in the standard theory, and he thus concluded that one must stipulate exactly when Process 1 occurs in order for quantum mechanics to be complete. Further, he argued that quantum mechanics together with a principle of charity about other observers requires an explicit mind–body dualism.

Wigner believed that he clearly had 'direct knowledge' of his own sensations. And he adopted a principle of charity with respect to other observers: while 'there is no strict logical reason to believe that others have similar experiences', he said, 'everybody believes that the phenomenon of sensations is widely shared by organisms that we consider to be living' (Wigner 1961: 175). Consequently, he felt that one must be able to tell some sort of quantum-mechanical story about how living beings come to have the determinate experiences they do.

Suppose that a system S (perhaps a patch on a phosphorescent screen) flashes in state ψ_1 and does not flash in state ψ_2. Wigner supposes that a friend F has the following two properties (which are essentially the same two dispositions as described above and earlier): (1) if the state of S is ψ_1, then the state of $F + S$ will be $\psi_1 \times \chi_1$ after the interaction where F observes S and (2) if the state of S is ψ_2, then the state of $F + S$ will be $\psi_2 \times \chi_2$ after the observation, and if the state of the observer is χ_1, then he answers the question 'Have you seen a flash?' with 'Yes'; and if the state of the observer is χ_2, then he answers this question with 'No'.

Wigner then considers what would happen if the state of the system S were in the linear combination $\alpha\psi_1 + \beta\psi_2$ of the two states ψ_1 and ψ_2 (Fig. 2.4): 'It *follows* from the linear nature of the quantum mechanical equations of motion that the state of the object plus observer is, after the interaction, $\alpha(\psi_1 \times \chi_1) + \beta(\psi_2 \times \chi_2)$. If we now ask the observer whether he saw a flash, he will with probability $|\alpha|^2$ say that he did' (and with probability $|\beta|^2$ say that he did not); and Wigner points out that, whatever answer his friend gives, he must be able to observe S himself and get

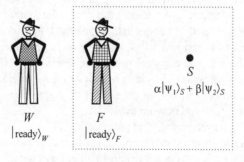

FIG. 2.4 Wigner's-friend setup.

the same answer, and so far quantum mechanics encounters no special difficulties: 'All this is quite satisfactory: the theory of measurement, direct or indirect, is logically consistent so long as I maintain my privileged position as ultimate observer' (Wigner 1961: 176).

But now consider the experiment from the perspective of Wigner's friend:

if having completed the whole experiment I ask my friend, 'What did you feel about the flash before I asked you?' He will answer, 'I told you already, I did [did not] see a flash,' as the case may be. In other words, the question whether he did or did not see a flash was already in his mind, before I asked him. If we accept this, then we are driven to the conclusion that the proper wave function immediately after the interaction of friend and object was already either $\psi_1 \times \chi_1$ or $\psi_2 \times \chi_2$ and not the linear combination $\alpha(\psi_1 \times \chi_1) + \beta(\psi_2 \times \chi_2)$. This is a contradiction, because the state described by the wave function $\alpha(\psi_1 \times \chi_1) + \beta(\psi_2 \times \chi_2)$ describes a state that has properties which neither $\psi_1 \times \chi_1$ nor $\psi_2 \times \chi_2$ has. If we substitute for 'friend' some simple physical apparatus, such as an atom, which may or may not be excited by the light-flash, this difference has observable effects and *there is no doubt that* $\alpha(\psi_1 \times \chi_1) + \beta(\psi_2 \times \chi_2)$ *describes the properties of the joint system correctly, the assumption that the wave function is either* $\psi_1 \times \chi_1$ *or* $\psi_2 \times \chi_2$ *does not.* If the atom is replaced by a conscious being, the wave function $\alpha(\psi_1 \times \chi_1) + \beta(\psi_2 \times \chi_2)$ (which also follows from the linearity of the equations) appears absurd because it implies that my friend was in suspended animation before he answered my question. (Wigner 1961: 176–7)

That is, if we believe the friend when he tells us that he had a determinate result to his observation even before we asked him what it was, and by his principle of charity Wigner says that we should, then we have to deny that the state $F + S$ was the superposition of the two determinate result states before we asked. Rather, $F + S$ must have been in a determinate result state, either $\psi_1 \times \chi_1$ or $\psi_2 \times \chi_2$, before we asked, which means that the collapse of the state must have occurred before we asked. But when exactly did it occur? Wigner believes that it must have occurred when the friend made his measurement because we would expect any other physical interaction to be correctly described by the linear dynamics. But then what is it about the friend that makes him different from other physical systems? The friend is a sentient being, and it is this that Wigner takes to justify treating him differently from inanimate physical systems— Wigner concludes, then, that it is by dint of the friend being *conscious* that he causes a collapse of the quantum-mechanical state of his object system.

Wigner argues that 'it follows that the being with a consciousness must have a different role in quantum mechanics than the inanimate measuring device ... In particular, the quantum mechanical equation of motion cannot be linear if the preceding argument is accepted' (Wigner 1961: 177). One could avoid this conclusion by denying that one's friend has determinate impressions and sensations, but 'to deny the existence of the consciousness of a friend to this extent is surely an unnatural attitude, approaching solipsism, and few people, in their hearts, will go along with it' (177–8).

Since the pure state $\alpha(\psi_1 \times \chi_1) + \beta(\psi_2 \times \chi_2)$ has different empirical properties from a statistical mixture of $\psi_1 \times \chi_1$ and $\psi_2 \times \chi_2$, Wigner recognized that it was a consequence of his position that one could, at least in principle, perform experiments to determine what systems are conscious and what are not (1961: 180–1). He noted, however, that David Bohm had shown that 'if the system is sufficiently complicated, it may be in practice impossible to ascertain the difference between certain mixtures, and some pure states' (Wigner 1961: 180; Bohm 1951, sects. 22.11, 8.27, and 8.28). That is, such experiments would typically be very difficult to perform. The reason that such experiments would be very difficult, perhaps even technologically impossible, has to do with the fact that interactions with the system's environment would typically (and very quickly) destroy the simple interference effects one might set out to measure. I shall discuss such decoherence effects in considerable detail later (in Chapter 8).

2.7   *The measurement problem (again)*

The collapse postulate is supposed to explain our determinate experience by guaranteeing the existence of determinate physical records at the end of our measurements. Von Neumann argued that it was unnecessary to say exactly how or when the collapse occurs. But Wigner recognized that it does in fact matter when the collapse occurs; it matters because we do not have a complete causal story without knowing this and because when the collapse occurs can have empirical consequences of a sort that von Neumann did not consider.

The measurement problem, then, might be put this way: The collapse dynamics is supposed to tell us what happens when a measurement is made and the usual deterministic dynamics is supposed to tell us what happens the rest of the time. If we suppose that measuring devices are ordinary physical systems interacting in their usual deterministic way, then we get a straightforward logical contradiction between the predictions of the two dynamical laws; so there must be something special about measuring devices, something that causes them to behave in a radically different way from other systems. But until we say exactly what it is that makes a measuring device behave differently, we do not have an empirically complete physical theory—we can describe experiments where the theory makes no empirical predictions because it is unclear which dynamical law one should use.

Wigner tried to solve the measurement problem by saying precisely what it is that distinguishes observers from all other physical systems: observers are *conscious*. But when you think about it, this does not really provide a very clear rule for when collapses occur. Would a cat cause a collapse? A mollusc? A rhododendron? Further, is it really necessary to introduce something *extra-physical* (in this case minds) in order to solve the measurement problem?!

3

THE THEORY OF THE UNIVERSAL
WAVE FUNCTION

E V E R E T T ' S description of the theory of the universal wave function, his relative-state formulation of quantum mechanics, is contained in three works: his thesis 'On the Foundations of Quantum Mechanics' (submitted to Princeton University on 1 March 1957), a short paper entitled '"Relative State" Formulation of Quantum Mechanics' (1957b), and a long paper 'The Theory of the Universal Wave Function' (1973). The main argument of the three works is the same, and many passages are very similar, sometimes identical. This chapter contains a close reading of the three works. The aim is to make clear what Everett himself actually said and why it is so difficult to figure out exactly what he had in mind.

At first sight Everett's proposal is simple enough: he wants to drop the collapse dynamics from the standard von Neumann–Dirac formulation of quantum mechanics, then deduce the empirical predictions of the standard theory as subjective appearances of observers who are themselves treated within pure wave mechanics as perfectly ordinary physical systems. The problem, however, is that it is unclear precisely how Everett intended to account for the determinate records and experiences of observers. As Richard Healey once put, Everett's interpretation of quantum mechanics itself stands in need of an interpretation (1984: 591).

3.1   What's wrong with von Neumann's theory?

Everett always begins the presentation of his own theory by first describing von Neumann's formulation of quantum mechanics. He explains that the wave function ψ is supposed to provide a complete and objective characterization of an isolated system (1973: 3). And he describes the two fundamentally different ways the state can change over time on the standard theory: Process 1 describes the discontinuous, random change brought about by observation, and Process 2 describes the continuous, deterministic change of an isolated system described by the linear Schrödinger dynamics (1973: 3).

In his thesis Everett asks the reader to consider an isolated system consisting of an observer plus an object system.

Can the change with time of the state of the *total* system be described by Process 2? If so, then it would appear that no discontinuous probabilistic process like Process 1 can take place. If not, we are forced to admit that systems which contain observers are not subject to the same kind of quantum mechanical description as we admit for all other physical systems. (1957*a*: 6)

If one opts for the first alternative, then the standard theory is logically inconsistent. If one opts for the second, then it is incomplete since it does not tell us what it takes to be an observer.

In his long paper Everett explicitly worries about the consistency and completeness of the standard formulation of quantum mechanics.

The question of the consistency of [von Neumann's formulation of quantum mechanics] arises if one contemplates regarding the observer and his object-system as a single (composite) physical system. Indeed, the situation becomes quite paradoxical if we allow for the existence of more than one observer.

In order to illustrate the paradoxical nature of the standard theory, Everett tells a Wigner's-friend-type story where an observer A measures the state of a system S, then the composite system of $A + S$ is observed by another observer B.

If we are to deny the possibility of B's use of a quantum mechanical description (wave function obeying the wave equation) for $A + S$, then we must be supplied with some alternative description for systems which contain observers (or measuring apparatus). Furthermore, we would have to have a criterion for telling precisely what type of systems would have the preferred positions of 'measuring apparatus' or 'observer' and be subject to the alternate description. Such a criterion is probably not capable of a rigorous formulation. (1973: 4)

Everett then abandons the hope that one can complete the standard formulation of quantum mechanics by providing a criterion for what constitutes a measurement. That is, he does not believe that anything like Wigner's proposed solution to the measurement problem will work.

Everett notes that one might naturally suppose that B can use the usual linear dynamics to describe the interaction between A and S, but he recognizes that this leads to problems given the standard way that one calculates quantum probabilities:

if we do allow B to give a quantum description to $A + S$, by assigning a state function ψ^{A+S}, then, so long as B does not interact with $A + S$, its state changes causally according to Process 2, *even though A may be preforming measurements upon S*. From *B*'s point of view, nothing resembling Process 1 can occur (there are no discontinuities), and the question of the validity of *A*'s use of Process 1 is raised. That is, *apparently* either *A* is incorrect in assuming Process 1, with its probabilistic implications, to apply to his measurements, or else *B*'s state function, with its purely causal character, is an inadequate description of what is happening to *A* + *S*. (1973: 4)

He concludes that

It is now clear that the interpretation of quantum mechanics with which we began is untenable if we are to consider a universe containing more than one observer. We must therefore seek a suitable modification of this scheme, or an entirely different system of interpretation. (1973: 6)

Everett thought of the measurement problem as one of several closely related problems faced by the standard collapse theory. Another problem was that the standard theory cannot account for *approximate* measurements.

In his thesis Everett asks 'What mixture of Processes 1 and 2 of the conventional formulation is to be applied to the case where only an approximate measurement is affected; that is, where an apparatus or observer interacts only weakly and for a limited time with an object system?' (1957*a*: 6). A satisfactory theory of approximate measurement must specify the probability for getting a particular reading of the measuring apparatus and tell us what the corresponding new state for the object system is. Everett, however, argues that while some approximate measurements could be treated using projection operators, it is impossible to treat all approximate measurements using von Neumann's method (1957*a*: 6–7, 13–15).[1]

Another problem concerns finding a satisfactory quantum theory of gravitation. Everett argues that quantizing general relativity raises 'serious questions about the meaning and interpretation of quantum mechanics when applied to so fundamental a structure as the space-time geometry itself' (1957*b*: 315). He says that his aim is to present a reformulation of quantum mechanics that can be applied to general relativity. Again, this requires a formulation of quantum mechanics that can be applied to the entire universe, and because of the measurement problem, the standard von Neumann formulation of quantum mechanics is inappropriate for

---

[1] More specifically, he shows that the relative system states for the measuring apparatus after an imperfect measurement typically cannot be generated from the original object system state by the application of any projection operator.

describing events in a closed and isolated system that contains observers (and the universe is presumably just such a system) (1957*b*: 315–16).

Everett saw such specific problems as different aspects of the same general problem: the standard formulation of quantum mechanics can only be consistently applied to systems that are subject to *external* observation:

The probabilities of the various possible outcomes of the observation are pre-scribed exclusively by Process 1. Without that part of the formalism there is no means whatever to ascribe a physical interpretation to the conventional machinery. But Process 1 is out of the question for systems not subject to external observation. (1957*a*: 8)

What Everett wanted was a formulation of quantum mechanics that could be understood as a complete and accurate physical description of any physical system, regardless of whether it was open or closed. Such a theory could be understood as providing a framework for a complete and accurate quantum-mechanical description of the entire universe, including the observers.

3.2   *Other formulations of quantum mechanics and their problems*

In his long paper Everett described five alternatives for developing a sat-isfactory formulation of quantum mechanics, one that can be consistently applied to closed systems.

Alternative 1 is just the standard collapse formulation of quantum mechanics but with the stipulation that there is in fact only one real observer (that is, I suppose that only *my* observations cause collapses). Everett says that 'this view is quite consistent, but one must feel uneasy when, for example, writing textbooks on quantum mechanics, describing Process 1, for the consumption of other persons to whom it does not apply' (1973: 6). But it is also difficult for me to see how this view is any better than the standard formulation of quantum mechanics since it is still unclear precisely when the collapse is supposed to occur. Does it occur when my retina becomes correlated with the pointer on my measuring device, or does it occur when my brain becomes correlated with my retina, or does it somehow occur when I become *aware* of the position of the pointer, etc.? And how one answers such questions here can make an empirical difference for the results of *self-measurements* of Wigner's-friend-type interference observables on such a theory.

Alternative 2 limits the applicability of quantum mechanics to exclude observers, measuring devices, or systems approaching macroscopic size.

One problem with this is that if we try to limit the applicability of the theory to exclude measuring devices or macroscopic systems, then we are faced with the difficulty of precisely defining the domain of the theory (1973: 6). Everett also appeals to von Neumann's principle of psychophysical parallelism to argue that giving special treatment to sentient beings is unacceptable, which, of course, also argues against Wigner's proposal:

And to draw the line at human or animal observers, i.e., to assume that all mechanical aparata obey the usual laws, but that they are somehow not valid for living observers, does violence to the so-called principle of psycho-physical parallelism, and constitutes a view to be avoided, if possible. To do justice to this principle we must insist that we be able to conceive of mechanical devices (such as servomechanisms) obeying natural laws, which we would be willing to call observers. (1973: 6–7)

Von Neumann's principle of psychophysical parallelism, which he in turn attributed to Bohr, was that it must be possible to describe the extraphysical process of subjective perception as if it were in reality in the physical world, to assign to its parts equivalent physical processes in the objective environment, in ordinary space (von Neumann 1955: 418). Everett's interpretation of the principle was that a satisfactory formulation of quantum mechanics must allow one to model an observer as an ordinary physical measuring device equipped with memory registers with which to record the results of its measurements. The idea is that if one is able to explain the states of such measurement records, then one could, at least in principle, explain the mental states of a real observer who made similar observations. Note that, while von Neumann used the principle of psychophysical parallelism to explain how the standard collapse theory works, Everett thought that this same principle was in fact incompatible with the standard collapse theory. Everett thought that a satisfactory formulation of quantum mechanics should allow one to explain precisely what happens when a measurement occurs and why one gets the result that one does, and, as we have already seen, the standard theory does not.[2]

Alternative 3 denies that an observer like B (in the Wigner's-friend-type experiment described in the last section) could ever be in possession of the state function of a composite system like $A + S$ (a system that contains a sentient observer!) because determining the state of such a system would require such a dramatic intervention that A would cease

---

[2] In some formulations of quantum mechanics, as we shall see later, the explanation of our determinate measurement records is not a purely physical one (many-minds theories, etc.). These theories violate the principle of psychophysical parallelism.

to function as an observer (and one might add that such an experiment would be virtually impossible to perform because of the complexity of the systems involved and the difficulty in isolating such systems from their environments—I shall discuss such issues in Chapter 8). Everett has two objections to this line of argument. First, regardless of what the state of $A + S$ is, quantum mechanics guarantees that it is in an eigenstate of some complete set of commuting operators and that measuring any of these will in no way disrupt the operation of A. In other words, the standard theory is perfectly compatible with our determining what the quantum-mechanical state of any system is without disturbing it. Secondly, Everett argues that it is irrelevant whether or not B knows the state of $A + S$; if he believes that it has an objective state that evolves deterministically, then he will also believe that it cannot be a random collapse that ultimately determines A's measurement result (1973: 7).

Alternative 4 abandons the state function as providing a complete physical description of a system. A might make his measurement and get a determinate result and B might still take the state function to provide a correct, though incomplete (since it would not tell him what result A got), description of the state of $A + S$. Everett says that the proponents of such an approach assume that one could ultimately get a deterministic theory where the quantum probabilities resulted from our ignorance concerning the values of the extra parameters that would supplement the state function to provide a complete state description. He does not say anything against such an approach at this point, but after his presentation of pure wave mechanics in his long paper Everett returns to his discussion of alternative formulations of quantum mechanics and describes Bohm's hidden-variable theory and what he does not like about it.

According to Bohm's theory each particle always has a determinate position and follows a continuous and deterministic trajectory and the wave function always evolves in the usual linear, deterministic way. One can think of the positions of an N particle system as being represented by a single point in $3N$-dimensional configuration space (since there are N particles and three position coordinates for each particle). One might then picture the motions of the particles by considering how that point would be pushed around by the probability current in configuration space as the wave function evolves. Everett's only explicit criticism of this theory is on the grounds of simplicity: 'if one desires to hold the view that ψ is a real field then the associated particle is superfluous since, as we have endeavored to illustrate, the pure wave theory is itself satisfactory' (1973: 112). In other words, Everett believes that he can account for our experience by appealing to the wave function alone. If this is right, then hidden variables like the particle positions in Bohm's theory are unnecessary and redundant. But this is all contingent on Everett being able to deliver a satisfactory account of our determinate experience on the basis of the deterministic evolution of the wave function alone.

Alternative 5, the one Everett likes, assumes 'the universal validity of the quantum description, by the complete abandonment of Process 1'. On this view

> The general validity of pure wave mechanics, *without any statistical assertions*, is assumed for *all* physical systems, including observers and measuring apparata. Observation processes are to be described completely by the state function of the composite system which includes the observer and his object-system, and which at all times obeys the wave equation (Process 2). (1973: 8)

On this approach the quantum-mechanical state is taken to provide a complete and accurate description of the physical state of the entire universe and the usual deterministic linear dynamics is assumed to provide a complete and accurate description of the time-evolution of this state.

Everett argues that this alternative has many advantages over the others considered. It is simple. It is applicable to the entire universe. All physical processes are treated equally, and in particular, measurements play no special role in the theory, which means, Everett argues, that the principle of psychophysical parallelism is automatically satisfied (1973: 8). On this view, one thinks of the wave function of the whole universe as the fundamental physical entity. Everett suggests calling it the theory of the universal wave function, 'since all of physics is presumed to follow from this function alone' (1973: 9).

Everett thought that his view corresponded closely with that held by Schrödinger, but he argued that such a view can only make sense when observation processes themselves are treated within the theory:

> It is only in this manner that the *apparent* existence of definite macroscopic objects as well as localized phenomena, such as tracks in cloud chambers, can be satisfactorily explained in a wave theory where the waves are continually diffusing. With the deduction in this theory that phenomena will appear to observers to be subject to Process 1, Heisenberg's criticism of Schrödinger's opinion—that continuous wave mechanics could not seem to explain the discontinuities which are everywhere observed—is effectively met. (1973: 115)

Everett wanted to show that observers, when treated within pure wave mechanics, are subject to something like an illusion where physical processes that are perfectly continuous and deterministic *seem* to be discontinuous and random and where properties that are in fact indeterminate, even indeterminate properties of macroscopic objects, appear to be determinate.

The relationship between Everett's view and Schrödinger's is interesting. As we have seen, Schrödinger found the idea of the collapse of the wave function objectionable, and he believed that the linear wave equation provided a complete and accurate description of the time-evolution of the state of the world. He initially thought that the wave function described something like the distribution of matter, and it may have been this line of thought that developed into his later position that the wave function represented 'simultaneous happenings'. Schrödinger said that

Nearly every result [a quantum theorist] pronounces is about the probability of this *or* that ... happening—with usually a great many alternatives. The idea that they be not alternatives but *all* really happen simultaneously seems lunatic to him, just *impossible*. He thinks that if the laws of nature took *this* form for, let me say, a quarter of an hour, we should find our surroundings rapidly turning into a quagmire, or sort of featureless jelly or plasma, all contours becoming blurred, we ourselves probably becoming jelly fish. It is strange that he should believe this. For I understand he grants that unobserved nature does behave this way—namely according to the wave equation. (Schrödinger 1995: 19; quoted in Lockwood 1996: 165)

He further argued that

The compulsion to replace *simultaneous* happenings, as indicated directly by the theory, by *alternatives*, of which the theory is supposed to indicate the respective *probabilities*, arises from the conviction that what we really observe are particles ... Once we have decided this, we have no choice. But it is a strange decision. (Schrödinger 1995: 20; see Lockwood 1996: 165)

In light of the similarity in their general views, it is easy to see why Everett acknowledged a debt to Schrödinger.

3.3 *Everett's project*

After describing the virtues of pure wave mechanics, Everett admitted that 'There remains, however, the question whether or not such a theory can be put into correspondence with our experience' (1973: 9). For his part, Everett was confident that the theory of the universal wave function could explain our determinate experience: '*The present thesis is devoted to showing that this concept of a universal wave mechanics, together with the necessary correlation machinery for its interpretation, forms a logically self consistent description of a universe in which several observers are at work*' (1973: 9). An important component of his interpretation of pure wave mechanics was his model of observers as physical measuring devices with registers for recording their measurement results. Everett described his main argument in his long paper as follows:

We shall be able to introduce into [the theory of the universal wave function] systems which represent observers. Such systems can be conceived as automatically functioning machines (servomechanisms) possessing recording devices (memory) and which are capable of responding to their environment. The behavior of these observers shall always be treated within the framework of wave mechanics. Furthermore, we shall deduce the probabilistic assertions of Process 1 as *subjective* appearances to such observers, thus placing the theory in correspondence with experience. We are then led to the novel situation in which the formal theory is objectively continuous and causal, while subjectively discontinuous and probabilistic. While this point of view thus shall ultimately justify our use of the statistical assertions of the orthodox view, it enables us to do so in a logically consistent manner, allowing for the existence of other observers. (1973: 9)

At the beginning of his short paper Everett describes his project as follows:

This paper proposes to regard pure wave mechanics (Process 2 only) as a complete theory. It postulates that a wave function that obeys a linear wave equation everywhere and at all times supplies a complete mathematical model for every isolated physical system without exception. It further postulates that every system that is subject to external observation can be regarded as part of a larger isolated system. (1957*b*: 316)

The idea is to start with no explicit interpretation of the wave function, explore the structure of pure wave mechanics, then choose an interpretation of the wave function based on that structure that accounts for our experience: 'The wave function is taken as the basic physical entity with *no a priori interpretation*. Interpretation only comes *after* an investigation of the logical structure of the theory. Here as always the theory itself sets the framework for its interpretation' (1957*b*: 316). And Everett explains the relationship between pure wave mechanics and the conventional formulation: 'The aim is not to deny or contradict the conventional formulation of quantum theory, which has demonstrated its usefulness in an overwhelming variety of problems, but rather to supply a new, more general and complete formulation, from which the conventional interpretation can be *deduced*' (1957*b*: 315). Indeed, the new theory is very much like the old one, only without the collapse postulate:

The new theory is not based on any radical departure from the conventional one. The special postulates in the old theory which deal with observation are omitted in the new theory. The altered theory thereby acquires a new character. It has to be analyzed in and for itself before any identification becomes possible between the quantities of the theory and the properties of the world of experience. The identification, when made, leads back to the omitted postulates of the conventional theory that deal with observation, but in a manner which clarifies their role and logical position. (1957*b*: 315)

Again the collapse postulate (Process 1) would be recaptured as descriptive of the *subjective experience* of observers, who are treated as ordinary physical systems within pure wave mechanics. In his thesis abstract Everett says:

The new theory results from the conventional formulation by *omitting* the special postulates concerned with external observation. In their place a concept of 'relativity of states' is developed for treating and interpreting the quantum description of isolated systems within which observation processes can occur. Abstract models for observers are formulated that can be treated within the theory as physical systems subject at all times to the same laws as all other physical systems. Isolated systems containing these model observers in interaction with other subsystems are investigated, and certain changes that occur in an observer as a consequence of the interaction with the surrounding systems are deduced. When these changes are interpreted as the experience of the observer this experience is found to be in accord with the statistical predictions of the conventional 'external observation' formulation of quantum mechanics. (1957*a*: 2)

That is, the physical changes that occur in the physical system representing an observer tell us what the observer experienced.

In summary, then, Everett's project was to assume (1) that a single wave function, an element of a Hilbert space, provides a complete and accurate description of the state of the entire universe and (2) that this universal wave function under all circumstances evolves according to the deterministic linear dynamics; and then from these two assumptions to deduce the statistical predictions of the standard formulation of quantum mechanics *as the subjective appearances of observers modelled as physical systems*. But while this may seem perfectly clear, it is difficult to figure out how his theory was supposed to work.

3.4   *The fundamental relativity of states*

Everett sometimes called his theory the relative-state formulation of quantum mechanics, and he insisted that it was the principle of the relativity of states that allowed one to interpret the quantum description of isolated systems. While the principle of the relativity of states is perfectly straightforward, it is difficult to determine the role that it is supposed to play for Everett in explaining our determinate experience.

If a system is described by a wave function, then there is generally no guarantee that each subsystem can be described by a wave function. Suppose, for example, that M measures the x-spin of an electron S in an eigenstate of z-spin. Given the dispositions of a good measuring device, the composite system will end up in a state like

$$\frac{1}{\sqrt{2}}(|x\text{-spin up}\rangle_M|\uparrow_x\rangle_S + |x\text{-spin down}\rangle_M|\downarrow_x\rangle_S), \qquad (3.1)$$

which is a nonseparable state where neither M nor S possesses an independent wave function of its own.

Concerning this sort of state Everett says that we notice that 'there is no longer any definite independent apparatus state, nor any independent system state. The apparatus therefore does not indicate any definite object-system value, and nothing like Process 1 has occurred' (1957a: 13). So how are we supposed to account for the fact that M seems to have a perfectly determinate record? After all, if a good observer measures the x-spin of a system, he will presumably end up getting the determinate result x-spin up or the determinate result x-spin down. The standard formulation of quantum mechanics predicts that the probability of each result will be $1/2$ in this situation, and this really does seem to describe what we observe. How then are we supposed to deduce *this* as a subjective appearance from pure wave mechanics when pure wave mechanics predicts a post-measurement state like the entangled superposition above?

Everett says that 'In order to bring about this correspondence with experience for the pure wave mechanical theory, we shall exploit the correlation between subsystems of a composite system that is completely described by a state function' (1973: 9). He explains that while a subsystem typically does not have a determinate state of its own, one can always ascribe to it a state *relative* to a specification of the state of the rest of the composite system. In the above post-measurement state, for example, while M did not get a determinate result to its measurement (indeed, M does not even have a determinate state of its own), the state $|x\text{-spin up}\rangle_M$ is uniquely picked out in the entangled superposition as the state of M *relative to* $|\uparrow\rangle_S$ being the state of S. It is not that $|x\text{-spin up}\rangle_M$ is the state of M and $|\uparrow\rangle_S$ is the state of S. Rather, given that the superposition describes the state of the composite system $M + S$, $|x\text{-spin up}\rangle_M$ is the state of M *relative to* $|\uparrow\rangle_S$ being the state of S, and $|x\text{-spin down}\rangle_M$ is the state of M *relative to* $|\downarrow\rangle_S$ being the state of S, etc.

One might think of *relative to*, then, as a relation between possible pure states of subsystems. Consider a composite system $S = S_1 + S_2$ in the state ψ_S. To every state ψ_{S_1} of S_1 we can associate a state of S_2, which is the state of S_2 relative to ψ_{S_1}—this relative state is written as $\psi_{S_2 \, \mathrm{rel} \psi_{S_1}}$. One can determine what $\psi_{S_2 \, \mathrm{rel} \psi_{S_1}}$ is by writing ψ_S in any basis where ψ_{S_1} occurs in exactly one term in the expansion. This term will have the form: $\psi_{S_1} \psi_{S_2 \, \mathrm{rel} \psi_{S_1}}$.[3]

Everett explains the fundamental relativity of states as follows:

Summarizing: There does not, in general, exist anything like a single state for one subsystem of a composite system. Subsystems do not possess states that are independent of the states of the remainder of the system, so that the subsystem states are generally correlated with one another. One can arbitrarily choose a state for one subsystem, and be led to the relative state for the remainder. Thus we are faced with a fundamental relativity of states, which is implied by the formalism of the composite systems. It is meaningless to ask the absolute state of a subsystem—one can only ask the state relative to a given state of the remainder of the subsystem. (1957b: 317)

The fact that Everett put this passage in italics is a sure sign of its significance, but how does the notion of relative states help to account for subjective appearances (which, after all, is Everett's goal)? In the conclusion of his long paper, Everett says

The 'quantum jumps' exist in our theory as *relative* phenomena (i.e., the states of an object-system relative to chosen observer states shows this effect), while the absolute states change quite continuously. (1973: 115)

After a typical measurement interaction the composite system is left in a superposition of states, but, in each of the superposed states (if one chooses an appropriate basis in which to write the state of the composite system) the measuring apparatus is described as having recorded a definite result and the object system is left in approximately an eigenstate

---

[3] See Everett's definition (1973: 38).

corresponding to that result. But the state of the composite system is the entangle superposition, not one or another of the relative states.

The discontinuous 'jump' into an eigenstate is thus only a relative proposition, dependent on the mode of decomposition of the total wave function into the superposition, and relative to a particular chosen apparatus-coordinate value. So far as the complete theory is concerned all elements of the superposition exist simultaneously, and the entire process is quite continuous. (1957*b*: 318)

The point here is that the jump into a state where there is a determinate measurement record is *only* a relative proposition—relative to a specification of the state of the rest of the composite system. While Everett almost always writes the post-measurement state of an observer and his object system in a representation where the observer can be said to have a definite record in each element of the superposition, he also describes one's choice of basis as *an arbitrary choice*—he insists that there is no preferred way to decompose the total quantum-mechanical state into relative states. Just as the post-measurement state can be written so that each term describes a state where the observer recorded a determinate result, it can also be written so that no term describes the observer as having recorded a determinate result. So not only is the result recorded by the observer only a relative proposition, but whether he recorded any determinate result whatsoever is only relative as well. But, then, how do these *relative* facts account for our determinate experience?

But perhaps we are asking this question too soon since *Everett does not take what has been said so far as explaining our determinate experience*. Rather, it is at precisely this point in his presentation, just after he describes the principle of the relativity of states and how one would apply the principle to a typical measurement situation, that he says that it is unclear how anyone *could* account for determinate measurement results in such a situation:

As a result of the interaction the state of the measuring apparatus is no longer capable of independent definition. It can be defined only *relative* to the state of the object system. In other words, there exists only a correlation between the states of the two systems. It seems as if nothing can ever be settled by such a measurement. (1957*b*: 318)

And he continues:

This indefinite behavior seems to be quite at variance with our observations, since physical objects always appear to us to have definite positions. Can we reconcile this feature [of] wave mechanical theory built purely on Process 2 with experience, or must the theory be abandoned as untenable? In order to answer this question we consider the problem of observation itself within the framework of the theory. (1957*b*: 318)

So instead of using it to provide an account of our determinate experience, Everett uses his most detailed discussion of the principle of the relativity of states to pose a problem for his theory: how can a global state where an observer has no determinate measurement record, but only various *relative* records, be understood as a state where the observer has the same determinate experience predicted by the standard formulation of quantum mechanics?

Everett sketches his answer to this question at the end of the introduction to his long paper. His answer seems to have something to do with the principle of the relativity of states, but it is unclear precisely what the relationship is:

Let one regard an observer as a subsystem of the composite system: observer + object-system. It is then an inescapable consequence that after the interaction has taken place there will not, generally, exist a single observer state. There will, however, be a superposition of the composite system states, each element of which contains a definite observer state and a definite relative object-system state. Furthermore, as we shall see, *each* of these relative object system states will be, approximately, the eigenstates of the observation corresponding to the value obtained by the observer which is described by the same element of the superposition. Thus, each element of the resulting superposition describes an observer who perceived a definite and generally different result, and to whom it appears that the object-system state has been transformed into the corresponding eigenstate. In this sense the usual assertions of Process 1 appear to hold on a subjective level to each observer described by an element of the superposition. We shall also see that correlation plays an important role in preserving consistency when several observers are present and allowed to interact with one another (to 'consult' one another) as well as with other object-systems. (1973: 10)

From this it seems that an observer's experience is to be explained by the fact that after a measurement there typically is *some* expansion of the state of the observer + object system where each element in the expansion is close to being an eigenstate of the observer getting a determinate result and the object system being in an eigenstate corresponding to that result. But how is *this fact* supposed to explain an observer's actual experience?

The global state of the composite system will typically not describe the observer as having recorded any particular determinate result. Moreover, almost all subsystem states that one might specify (in the Hilbert-space sense of *almost all*) would determine a relative state where the observer failed to record any determinate result. Everett knows this, but at the same time he insists that the choice of basis is arbitrary.

In fact, to any arbitrary choice of state for one subsystem there will correspond a *relative state* for the other subsystem, which will generally be dependent upon the choice of state for the first subsystem, so that the state of one subsystem is not independent, but correlated to the state of the remaining subsystem. (1973: 10)

In order to get a relative state where the observer records a determinate result, one must specify a very special state for the remainder of the system (or for the observer himself)—if the choice of basis is arbitrary, then what justifies this very special choice?

Everett's discussion of the principle of relative states leaves us with a problem: how does the fact that an observer's results are determinate in each element of a *very special* expansion of the global wave function account for the fact that it *appears to the observer* that he in fact recorded a single determinate measurement result?

3.5   *The appearance of phenomena*

After describing some of the properties of quantum mechanics without the collapse postulate (Process 1), Everett explains that

This... has the far reaching implication that for any possible measurement, for which the initial system state is not an eigenstate, the resulting state of the composite system leads to *no* definite system state nor any definite apparatus state. The system will not be put into one or another of its eigenstates with the apparatus indicating the corresponding value, and nothing resembling Process 1 can take place. (1973: 60)

It seems, then, that Everett cannot intend to *simply deduce* Process 1 from pure wave mechanics alone, nor can he claim that the pure wave mechanics by itself typically predicts the same post-measurement states as the standard collapse formulation of quantum mechanics.

Everett describes a typical measurement interaction and the resulting entangled superposition of the measuring apparatus and the object system:

Thus in general after a measurement has been performed there will be no definite system state nor any definite apparatus state, even though there is a correlation. It seems as though nothing can ever be settled by such a measurement. Furthermore this result is independent of the *size* of the apparatus, and remains true for apparatus of quite macroscopic dimensions. (1973: 61)

To illustrate this, Everett considers an experiment where one couples an x-spin measuring device to a cannonball so that if the particle is found to be x-spin up, then the cannonball will be shifted one foot to the left; and if it is found to be x-spin down, then the cannonball will be shifted an equal distance to the right. If the particle begins in a superposition of x-spin up and x-spin down, the linear dynamics requires that the cannonball will end up in an entangled superposition of two positions, which Everett takes to mean that 'There is no definite position for our macroscopic cannonball!' (1973: 61).[4]

Note that in Everett's discussion of the determinate properties of cannonballs, he wholeheartedly accepts the consequences of the standard eigenvalue–eigenstate link concerning the conditions under which a physical property is determinate. There is further evidence that he accepted the standard eigenvalue–eigenstate link in the longer quotations above. If he really did intend to maintain the standard interpretation of states in some sense, then this is a critically important point for understanding his formulation of quantum mechanics.[5]

But if macroscopic systems typically fail to have determinate positions and if observers' physical records are similarly indeterminate, then this is a serious problem for pure wave mechanics:

This behavior seems to be quite at variance with our observations, since macroscopic objects always appear to us to have definite positions. Can we reconcile this prediction of the purely wave mechanical theory with experience, or must we abandon it as untenable? (1973: 61–2)

And Everett again explains the problem and his strategy for solving it:

We saw ... that in general a measurement ... had the outcome that neither the system nor the apparatus had any definite state after the interaction—a result seemingly at variance with our experience. However, we do not do justice to the theory of pure wave mechanics until we have investigated what the theory itself has to say about the *appearance* of phenomena to observers, rather than hastily concluding that the theory must be incorrect because the actual states of systems as given by the theory seem to contradict our observations. (1973: 63)

---

[4] Given that he could have used almost any macroscopic system to make this point, that Everett chose to use a cannonball may go some way in explaining how he ended up working with the Weapons Systems Evaluation Group at the Pentagon after receiving his Ph.D. from Princeton (1957a: 1). Everett later worked at the Institute for Defense Analysis.

[5] That he assumes the eigenvalue–eigenstate link without comment in his discussion of determinate properties is perhaps the best evidence for David Albert's claim (1992: 116) that Everett may have had something like the bare theory in mind (Ch. 4).

While an observer's physical records are typically indeterminate, perhaps it really is possible to explain why they none the less *seem* to be perfectly determinate to an observer treated within the theory. It is, after all, not at all clear what it would be like *to be* an observer who had in fact recorded a superposition of results—what does the theory predict concerning such an observer's experience? Would it be anything like *our* experience?

But then again how *could* pure wave mechanics predict a single determinate appearance when appearances on Everett's theory are supposed to be fully determined by the states of physical records? If the linear dynamics is taken to describe the time-evolution of all systems at all times, then after a typical measurement neither an observer M (treated as a physical system) nor the object system will have a determinate state, but will both be part of an entangled superposition like

$$\frac{1}{\sqrt{2}}(|\text{recorded } x\text{-spin up}\rangle_M|\uparrow_x\rangle_S + |\text{recorded } x\text{-spin down}\rangle_M|\downarrow_x\rangle_S).$$

$$(3.2)$$

On the standard interpretation of states this state does not describe a situation where the observer recorded any single, determinate measurement result. Indeed, if one assumes the principle of state completeness (which Everett did), then it is difficult to see how such a state could describe a situation where the observer had a single determinate measurement record on *any* interpretation of the states—if this state is supposed to describe a situation where the observer recorded a single, determinate result, then which result does it describe him recording? Because of the symmetry between the two terms, this state cannot be taken to select either of the two possible results. Moreover, our empirical evidence presumably requires that any superposition of the two results be compatible with actually obtaining *either* result (because it is empirically possible for one to get any result that is associated with a nonzero amplitude!). It seems, then, that there can be nothing in the quantum state predicted by the usual linear dynamics alone that determines which measurement result the observer got. If states like this are supposed to be complete and accurate descriptions of the composite system $M + S$ (and Everett insists that they are), then pure wave mechanics is simply incompatible with the claim that an observer typically records a single, determinate measurement result.[6]

---

[6] If there is a single, determinate result, then the post-measurement state is incomplete since it does not specify what that result is. One could respond that the post-measurement state is complete in the sense that it completely determines the quantum-mechanical probabilities. This is right, but I think a little misleading. If there is a physical matter of fact

Compare how measurements work in pure wave mechanics and in the standard collapse formulation of quantum mechanics. On the standard theory the fact that an observer typically records a determinate, repeatable measurement result is explained by the collapse of the state on measurement, and the fact that our measurement results seem to be randomly distributed and have the relative frequencies that they do is explained by the *way* that the state collapses. On the standard theory, the composite system randomly ends up in an eigenstate of the observer recording one or other of the two possible measurement results: in the above experiment either $|x\text{-spin up}\rangle_M |\uparrow_x\rangle_S$ or $|x\text{-spin down}\rangle_M |\downarrow_x\rangle_S$, each with probability $1/2$. According to the standard interpretation of states, each of these are states where the observer has in fact recorded a perfectly determinate measurement result. And since one would presumably want any physical theory that one is taking seriously to predict that a post-measurement state will typically be one where an observer has a determinate record of the measurement outcome, this is good. But in pure wave mechanics an observer typically does not even have a determinate quantum-mechanical state of his own, let alone a state that describes him as having recorded a determinate measurement result, and this is bad.

Everett, however, does not respond to the determinate-experience problem by claiming that after a typical measurement an observer does in fact have, contrary to the standard interpretation of states, a determinate physical state and a determinate record of the outcome; rather, he argues that one must consider the *appearance* of phenomena to observers. An observer who is actually in an entangled superposition of having recorded mutually incompatible results and who thus has no determinate measurement record would none the less experience the *appearance* of a determinate result. On the other hand, we have just seen that there can be no single, determinate physical record represented in the usual quantum-mechanical state predicted by the linear dynamics.

This dilemma suggests that in order to make sense of Everett's position one might ultimately want to distinguish between *physical* states which evolve in a deterministic way but where there are typically no determinate records and *mental* states which represent the determinate subjective appearances experienced by observers and which evolve in a

---

about what result the observer recorded, then since the usual quantum-mechanical state does not tell us what the physical record is here, the usual quantum-mechanical state does not completely specify all physical matters of fact concerning the composite system and is thus not a complete physical description.

stochastic way.[7] But this sort of distinction between physical and mental states would be incompatible with how Everett understands observers and their mental states. Not only does he require that a satisfactory formulation of quantum mechanics satisfy the principle of psychophysical parallelism, but he took an observer's physical state to determine fully his mental state.

We are faced with the task of making deductions about the appearance of phenomena on a subjective level, to observers which are considered as purely physical systems and are treated within the theory. In order to accomplish this it is necessary to identify some objective properties of such an observer (states) with subjective knowledge (i.e., perceptions). (1973: 63)

An observer for Everett is modelled as an automaton. The observer's knowledge and experience are fully determined by the physical state of his memory registers. The physical state $\psi^O_{[A,B,...]}$ represents a situation where the observer O has determinately recorded in his memory the events $A, B, \ldots$ As Everett puts it, 'These configurations can be thought of as punches in a paper tape, impressions on a magnetic reel, configurations of a relay switching circuit, or even configurations of brain cells' (1973: 65).

The automaton model of observers provides Everett with a final description of his problem and ultimate goal:

Our problem, then, is to treat the interaction of such observer-systems with other physical systems (observations), within the framework of wave mechanics, and to deduce the resulting memory configurations, which we can then interpret as the subjective experiences of the observers. (1973: 65)

From this it might sound as if Everett intends to show that after an x-spin measurement, say, an observer will end up with precisely one of the determinate memory configurations predicted by the standard collapse formulation of quantum mechanics, either $\psi^O_{[\uparrow_x]}$ or $\psi^O_{[\downarrow_x]}$. But again, we have already seen that if the linear dynamics describes the measurement process and if one assumes the principle of state completeness, then this cannot be right. Further, this cannot be what Everett himself has in mind since he has already argued that nothing like Process 1 can occur in the time-evolution of the quantum-mechanical state (which, by the way, is simply a consequence of assuming that the deterministic dynamics always describes the time-evolution of the state). Moreover, he has argued that an

---

[7] This distinction would presumably lead to something like the single-mind or many-minds theories of Ch. 7.

observer has no determinate *physical* record after a typical measurement. If one supposes that a determinate physical record is necessary for there to be a determinate mental record, which seems to be a reasonable enough assumption about the relationship between physical and mental states, then there can be no determinate *mental* record either!

3.6 *The deduction of subjective appearances*

Everett begins his deduction of the subjective appearances predicted by the standard formulation of quantum mechanics by defining what it means for an observer to make a good observation. On his definition, an observation is *good* if and only if the interaction between the observer and the object system is such that if the object system begins in an eigenstate of the observable being measured, then its state would be unchanged and the memory configuration of the observer would end up in an eigenstate of recording a determinate result that represents which eigenstate the object system is in (1957*a*: 19, 1973: 65–6). 'The requirement that the eigenstates for the system be unchanged is necessary if the observation is to be significant (repeatable), and the requirement that the observer state change in a manner which is different for each eigenfunction is necessary if we are to be able to call the interaction a measurement at all' (1973: 66).

Note that Everett's automaton model of observers and his conditions for a good observation correspond to how we have been treating observers all along: (1) if a good observer measures an object system that is initially in an eigenstate of the observable being measured, then the observer and the object system will end up in a separable state where the observer has recorded which eigenstate the object system was in and the object system is still in that eigenstate; and it follows from this and the linear dynamics that (2) if a good observer measures an object system that is not in an eigenstate of the observable being measured, then the observer and object system will end up in a correlated superposition of the observer having recorded various incompatible results and the object system being in the corresponding eigenstates. Everett calls these Rule 1 and Rule 2, respectively (1973: 67). Rule 2 just says that Rule 1 can be applied separately to each element of a superposition and then superpose the results to obtain the final state after the measurement interaction. The fact that each element can be thought of as separately obeying the wave equation follows from the linearity of the dynamics (1973: 67). In general, then, if an observer M begins in a ready-to-make-a-measurement state and measures the observable O of system S, with eigenstates ϕ_S^i, then the total state transforms according to

$$|\text{ready}\rangle_M \sum_i a_i \phi_S^i \rightarrow \sum_i a_i |a_i\rangle_M \phi_S^i. \qquad (3.3)$$

And the job at hand is that 'We must now seek the interpretation of such final total states' (1973: 68).

The key to explaining the determinate perceptions of an observer is to be found by considering the post-measurement state *term by term* (in the observer's determinate-record basis).

We note that there is no longer any independent system state or observer state, although the two have become correlated in a one–one manner. However, in each *element* of the superposition . . . the object-system state is a particular eigenstate of the observer, and *furthermore the observer-system state describes the observer as definitely perceiving that particular system state*. It is this correlation that allows one to maintain the interpretation that a measurement has been performed. (1973: 68, 1957a: 23)

But we are still faced with the same question as before: how is *this fact* supposed to account for the observer's determinate experience? Everett marks the end of this passage with a long footnote:

At this point we encounter a language difficulty. Whereas before the observation we had a single observer state afterwards there were a number of different states for the observer, all occurring in a superposition. Each of these separate states is a state for an observer, so that we can speak of the different observers described by the different states. On the other hand, the same physical system is involved, and from this viewpoint it is the *same* observer, which is in different states for the different elements of the superposition (i.e., has had different experiences in the separate elements of the superposition). In this situation we shall use the singular when we wish to emphasize that a single physical system is involved, and the plural when we wish to emphasize the different experiences for the separate elements of the superposition. (1973: 68 n.)

And this is one of Everett's two most detailed explanations of how pure wave mechanics accounts for our determinate experience![8]

Everett insists that the observer is represented by a single physical system throughout the measurement. And he says that it is only with regard

---

[8] I shall discuss the other detailed explanation, the 'Note added in proof' to his 1957b paper, in Sect. 3.8.

to the observer's subjective experience that one would talk of there being many observers after the measurement. There is a sense, then, in which each *physical* observer has many mutually incompatible experiences and memories. But then how does one explain why it *seems* that one typically ends up with a *single* determinate measurement result? If there is only one physical me associated with many mutually incompatible experiences, then if I do in fact get a single determinate result, it must be because *my* mental state is not determined by my physical state. But then what does determine *my* mental state?

I might respond by denying that I do in fact typically get a single, determinate measurement result. In the spirit of the fundamental relativity of states one might deny that there is any absolute matter of fact about what measurement result I recorded and say that such facts are *relative facts*: I saw the pointer pointing to 'x-spin up' *relative to* my object system being x-spin up, but I saw the pointer pointing to 'x-spin down' *relative to* my object system being x-spin down. But since on this view there is no absolute matter of fact about what I saw, it is unclear exactly how this is supposed to account for my experience (experience about which there really does seem to me to be an absolute matter of fact). Or put somewhat differently, if what I experience is a relative fact, then what is it that explains why it does not seem to me that it is?![9]

Without further explanation, Everett says that we can now carry the discussion a step further by considering what would happen if M repeated his measurement on S. Given the properties of a good measuring device and the linearity of the dynamics, the state will evolve as follows:

$$|a_i, \text{ready}\rangle_M \sum_i a_i \phi_S^i \rightarrow \sum_i a_i |a_i, a_i\rangle_M \phi_S^i. \tag{3.4}$$

Everett concludes:

Again, we see that each element ... describes a system eigenstate, but this time also describes the observer as having obtained the *same result* for each of the two observations. Thus for every separate state of the observer in the final superposition, the result of the observation was repeatable, even though different for different states. This repeatability is, of course, a consequence of the fact that after an observation the *relative* system state for a particular observer state is the corresponding eigenstate. (1973: 69, 1957a: 23)

---

[9] I shall consider such puzzles, and possible resolutions, more in later chapters. Everett's insistence on the notion of relative state here strongly suggests something like the relative-fact reading that I consider in Ch. 7.

And that is Everett's explanation for why observers have the same subjective experiences when they repeat their measurements.

Everett also shows that two observers who make the same measurement (of a quantity like x-spin) on an undisturbed system and compare their results will end up in a state where each term of the determinate-record expansion describes both observers as having the same measurement record and the object system being in the corresponding eigenstate: 'This means that observers who have separately observed the same quantity will *always* agree with each other' (1973: 80).

Note here, however, that each of the last two results are contingent on the measurements being *perfect*. If one does not repeat a measurement in precisely the same way as one performed it the first time, then one will typically fail to record the same result in each term of the quantum-mechanical state. Similarly, if the two measuring devices do not interact with the object system in precisely the same way, then their records will not be perfectly correlated and there will be a nonzero amplitude associated with a state where their results *do not agree*. So Everett needs to say something about imperfect measurements. But I shall return to this after I discuss the experiments that are suppose to allow one to deduce the usual statistical predictions of quantum mechanics.

In order to deduce the usual quantum statistical predictions as subjective appearances, Everett considers what would happen if M measured the same observable O of a number of separate systems $S_1, S_2, \ldots$ that are all in the same initial state $\psi_{S_n} = \sum a_i |\phi_i\rangle_{S_n}$, where $|\phi_i\rangle$ are again eigenfunctions of A.

If the initial state of the composite system is

$$|\text{ready}_1, \text{ready}_2, \ldots\rangle_M \phi_{S_1} \phi_{S_2} \ldots, \tag{3.5}$$

then the state after M's first measurement will be

$$\sum_i a_i |a_i, \text{ready}_2, \ldots\rangle_M |\phi_i\rangle_{S_1} \phi_{S_2} \ldots \tag{3.6}$$

and the state after M's second measurement will be

$$\sum_{i,j} a_i a_j |a_i, a_j, \ldots\rangle_M |\phi_i\rangle_{S_1} |\phi_j\rangle_{S_2} \ldots \tag{3.7}$$

And so on, as M's memory registers become correlated with the states of the systems being measured.

This ever more complicated total state is a superposition of states each of which describes the observer with a definite memory sequence and the systems that have been measured in corresponding eigenstates. Indeed, every possible sequence of measurement results is represented by some term in the final superposition. Concerning this state Everett concludes that

In the language of subjective experience, the observer which is described by a typical element ... of the superposition has perceived an apparently random sequence of definite results for the observations. It is furthermore true, since in each element the system has been left in an eigenstate of the measurement, that if at this stage a redetermination of an earlier system observation ... takes place, every element of the resulting final superposition will describe the observer with a memory configuration ... in which the earlier memory coincides with the later—i.e., the memory states are *correlated*. It will thus *appear* to the observer which is described by a typical element of the superposition that each initial observation on a system caused it to 'jump' into an eigenstate in a random fashion and thereafter remain there for subsequent measurements on the same system. Therefore, qualitatively, at least, the probabilistic assertions of Process 1 *appear* to be valid to the observer described by a typical element of the final superposition. (1973: 70)

And thus he claims that he has deduced the qualitative phenomena of the collapse of the wave function (Process 1) in terms of the subjective experiences of observers treated within pure wave mechanics—even though there is, as a matter of fact, never a collapse of the quantum-mechanical state.

In order to derive the quantitative assertions of Process 1 as well, Everett introduces a measure on the elements in the final superposition so that he can talk sensibly about what the statistical properties of the measurement results would be for a *typical* observer in the superposition. What he wants to show is that the usual quantum statistics would be exhibited by the sequence of results represented by a *typical* element in the final superposition (1973: 71). He then takes this together with the considerations above to constitute a complete deduction of the usual statistical predictions of quantum mechanics as subjective appearances of an observer treated within pure wave mechanics.

Everett argues that the only mathematically natural probability measure on the elements of the final superposition is the one determined by the norm-squared of the amplitude associated with each element. The measure of a set of elements, then, is given by the sum of the norm-squared of the amplitude of each term in the set (1973: 71–2). And it is with this measure that we are supposed to deduce the usual quantum statistics as appearances for a typical observer in the final superposition of the total state describing the corresponding physical observer.

If we consider the sequences to become longer and longer (more and more observations performed) *each* memory sequence of the final superposition will satisfy any given criterion for a randomly generated sequence, generated by the independent probabilities $a_i^* a_i$ [the norm-squared of the coefficient on each element of the object system state], except for a set of total measure which tends toward zero as the number of observations becomes unlimited. Hence all averages of functions over *any* memory sequence, including the special case of frequencies, can be computed from the probabilities $a_i^* a_i$, except for a set of memory sequences of measure zero. We have therefore shown that the statistical assertions of Process 1 will appear to be valid to *almost all* observers described by separate element of the [final] superposition ... in the limit as the number of observations goes to infinity. (1973: 74)

Everett then argues that this result can be generalized to apply to arbitrary sequences of observations (1973: 74–7).

Throughout the discussion Everett insists that all physical processes are 'entirely deterministic and continuous' and lead to a superposition 'each element of which describes the observer with a different memory state' and thus explains why it '*appears* to the observer that the probabilistic aspects of the usual form of quantum theory are valid': 'We have thus seen how pure wave mechanics, without any probability assertions, can lead to these notions on a subjective level, as appearances to observers' (1973: 78).

Everett presents essentially the same argument in his earlier paper (1957*b*). After describing the state that would result from performing the same measurement on a number of identically prepared systems, he notes that

A typical element ... of the final superposition describes a state of affairs wherein the observer has perceived an apparently random sequence of definite results for the observations. Furthermore the object systems have been left in the corresponding eigenstates of the observation. (1957*b*: 320)

And if the observer repeats a measurement, then each element in the final state will describe him as recording the same result as he did the first time he performed that measurement.

It will thus *appear* to the observer, as described by a typical element of the superposition, that each initial observation on a system caused the system to 'jump' into an eigenstate in a random fashion and thereafter remain there for subsequent measurements of the same system. Therefore—disregarding for the moment quantitative questions of relative frequencies—the probabilistic assertions of Process 1 *appear* to be valid to the observer described by a typical element of the final superposition. (1957*b*: 320)

Concerning quantitative questions of relative frequencies, Everett notes that as more and more observations are performed, '*each* memory sequence of the final superposition will satisfy any given criterion for a randomly generated sequence, generated by the independent probabilities [given by Process 1], except for a set of total measure which tends toward zero as the number of observations becomes unlimited' (1957*b*: 322). Everett thus concludes that 'We have therefore shown that the statistical assertions of Process 1 will appear to be valid to the observer, *in almost all* elements of the superposition . . . in the limit as the number of observations goes to infinity' (1957*b*: 322, 1957*a*: 31). And this means that 'all predictions of the usual theory will appear to be valid to the observer in almost all observer states' (1957*b*: 322, 1957*a*: 32).

It is important to be clear about the mathematical facts here since Everett's description allows for some confusion. It is typically false that most terms in the final superposition (written in the observer's determinate-record basis) in the ordinary counting sense of *most* will describe sequences of measurement results that have the right relative frequencies, but it is always true that most of the elements in the final superposition *in the norm-squared-coefficient measure* will describe sequences of results that have close to the right relative frequencies. So it is this second fact that must somehow explain why Process 1 will appear to be valid to a typical observer.[10]

A similar line of argument is also supposed to explain what happens when one repeats a measurement after an intervening measurement of an incompatible observable. If an observer measures one observable, say x-spin, and then measures a noncommuting observable of the same system, say z-spin, then the observer will not necessarily get the same result for a subsequent measurement of the first observable (x-spin) in every element of the total composite state. But if this *sequence* of measurements

---

[10] But again, all of the observers described by the superposition are the *same* observer in that they all correspond to the same physical system—it is just that this one physical observer somehow has many mutually incompatible experiences and memories. A typical memory sequence, in the norm-squared sense of typical, will exhibit the usual quantum statistics. But just as one might wonder how having many mutually incompatible experiences accounts for one's own determinate experience, one might wonder why it is the norm-squared measure that is relevant to one's experience. I shall discuss both of these problems later.

(sequence of three measurements here) is repeated on similarly prepared systems, then as the number of systems observed gets large almost all (in the norm-squared measure) of the terms in the determinate-record expansion of the final state will describe the observer as having recorded the same relative frequencies for the *sequence* of measurements as predicted by the standard formulation of quantum mechanics.

If such stories work at all, a similar story would provide a way of understanding imperfect measurements. While it is unlikely that any two observers, or even the same observer repeating a measurement, will ever perform exactly the same measurement on a system twice, it must be the appearance of the right joint relative frequencies in a typical element of the final superposition that explains the sort of repeatability of observations that we see in real experiments. If even a single observation is made on a system and is then followed by a similar observation of the same system, then most of the terms in the post-measurement state, in the norm-squared measure, not the usual counting measure, will describe situations where both observations yielded the same result. As a large number of such experiments are performed, most of the terms in the final state (in the norm-squared measure) will describe a situation where most experiments yielded the same result for both measurements.

Everett also thought that the EPR experiment was easily understood in the context of pure wave mechanics, and he claimed that 'one observer's observation upon one system of a correlated but non-interacting pair of systems, has no effect on the remote system, in the sense that the outcome or expected outcome of any experiments by another observer on the remote system are not affected' (1973: 83). But the extent to which this is true depends, as we shall see, on exactly how one understands pure wave mechanics and how it is supposed to account for our experience.[11]

In any case, Everett believed that he had deduced the standard statistical predictions of quantum mechanics (for the sorts of experiment we in fact perform), as subjective appearances, from a perfectly deterministic physical theory.

We have now completed the abstract treatment of measurement and observation, with the deduction that the statistical predictions of the usual form of quantum theory (Process 1) will appear to be valid to all observers. We have therefore succeeded in placing our theory in correspondence with experience, at least insofar as the ordinary theory correctly represents experience. (1973: 85)

---

[11] It is true, for example on the bare theory described in Ch. 4 but not on the revised single-mind theory described in Ch. 7.

He claimed that

The theory based on pure wave mechanics is a conceptually simple causal theory, which fully maintains the principle of the psycho-physical parallelism. It therefore forms a framework in which it is possible to discuss (in addition to ordinary phenomena) observation processes themselves, including the inter-relationships of several observers, in a logical, unambiguous fashion. (1973: 118)

And that

While our theory justifies the personal use of the probabilistic interpretation as an aid to making practical predictions, it forms a broader frame in which to understand the consistency of that interpretation. (1973: 118)

And he believed that this formulation of quantum mechanics may prove fruitful in field theories since 'one can assert that field equations are satisfied everywhere and everywhen, then *deduce* any statistical assertions by the present method' (1973: 119). It also 'avoids the necessity of considering anomalous probabilistic jumps scattered about space-time' (1973: 119), which, of course, would help one to reconcile quantum mechanics with relativity. And finally 'The wave theory is definitely tenable and forms, we believe, the simplest complete, self-consistent theory' (1973: 115).

Everett concludes by noting that

Aside from any possible practical advantages of the theory, it remains a matter of intellectual interest that the statistical assertions of the usual interpretation do not have the status of independent hypotheses, but are deducible (in the present sense) from the pure wave mechanics that starts completely free of statistical postulates. (1957*b*: 323, 1957*a*: 36)

Or put another way, even if one does not like the theory, one should find it interesting that the collapse dynamics is not independent of the linear dynamics. If this is true, then it is certainly interesting.

3.7   *The mechanics of macroscopic systems*

Everett took one of the virtues of his formulation of quantum mechanics to be that it explains the classical appearance and behaviour of macroscopic systems. He starts by considering an electron and a proton each with a definite momentum, in a box. If each particle has a definite momentum, then neither particle has a determinate position (which is precisely what the standard interpretation of states would say). One would expect a hydrogen atom in the ground state to form in the box eventually.

We notice, however, that the position amplitude density of each particle is *still* uniform over the whole box. Nevertheless the amplitude distributions are now no longer independent, but correlated. In particular, the *conditional* amplitude density for the electron, conditioned by any definite proton ... position, is *not* uniform but is given by the familiar ground state wave function for the hydrogen atom. What we mean by the statement, 'a hydrogen atom has formed in the box,' is just that this correlation has taken place—a correlation which insures that the *relative* configuration for the electron, for a definite proton position, conforms to the customary ground state configuration. (1973: 86)

So while the general state does not describe a hydrogen atom with a determinate position, Everett argues that it does describe a hydrogen atom, and that the same line of argument holds for more complex systems constructed by way of strong correlations (1973: 86–7). Everett notes that while it is a consequence of the linear dynamics that a cannonball would typically evolve to a state where its centre of mass had no determinate position, any specified centre of mass would pick out a state that describes a *cannonball*—that is, the particles that make up the cannonball do not spread out independently but rather preserve the structure of the macroscopic object and end up in a final state that can be described as a superposition of cannonballs at different positions (1973: 87–8).

The quantum-mechanical state of a system of macroscopic objects, a collection of cannonballs say, does not ascribe anything like determinate positions or momenta to the individual objects. But any such state 'can at any instant be analyzed into a *superposition* of states each of which *does* represent the bodies with fairly well defined positions and momenta' (1973: 89). For real macroscopic systems these local branch states are not states where the systems have determinate positions and momenta; rather they are states where the product of the uncertainties is very small by macroscopic standards. The idea, then, is that one could use a basis that provides such 'quasi-classical' branch states to account for our determinate experience of macroscopic objects.[12]

---

[12] The notion of a quasi-classical state will play a central role in how Gell-Mann and Hartle understand Everett's formulation of quantum mechanics later (Ch. 8). It is worth noting, however, that Everett's discussion of quasi-classical states here is not meant to explain why observers have *determinate* experiences. This is something that he believed that he had already explained in his short paper and in the first part of his long paper by noting that an observer will have a determinate record in each term of the global state (when written in an appropriate basis). Given that he had already explained why there are determinate

Each of these states then propagates approximately according to classical laws, so that the general state can be viewed as a superposition of quasi-classical states propagating according to nearly classical trajectories. In other words, if the masses are large or the time short, then there will be strong correlations between the initial states (approximate) positions and momenta and those at a later time, with the dependence being given approximately by classical mechanics. (1973: 89)

An observer will not see macroscopic objects as 'smeared out' over large regions of space and thus notice that the actual physical state is a super-position of very different quasi-classical states. Rather, when he makes his observation, the observer will become correlated with the macroscopic system in such a way that relative to his recording a (roughly) determinate result, the macroscopic system will be in the single corresponding quasi-classical state:

After the observation the composite system of objects + observer will be in a superposition of states, each element of which describes an observer who has perceived that the objects have nearly definite positions and momenta, and for whom the relative system state is a quasi-classical state ... and furthermore to whom the system will appear to behave according to classical mechanics if his observation is continued. We see, therefore, how the classical appearance of the macroscopic world to us can be explained in the wave theory. (1973: 90)

That is, there will be some expansion of the global state where each term describes observers as having almost determinate records of *both* the positions and momenta of macroscopic systems, and one would expect the values of these almost determinate quantities to evolve in an almost classical way. And Everett takes this to be enough to account for the classical appearance of such systems. There are, however, two natural questions: (1) how does the existence of such an expansion explain our determinate

---

experiences of macroscopic objects, his discussion of quasi-classical states was supposed to explain why one would judge that these determinate experiences agree with the predictions of *classical* mechanics. This distinction is important. If I am right, then environmental decoherence has nothing whatsoever to do with Everett's explanation for why observers have determinate experiences. Indeed, if I understand it correctly, Everett's notion of a quasi-classical state does not even rely on environmental decoherence; rather, it seems that he just wants the reader to think of the global state as a superposition of states describing macroscopic systems (the quasi-classical states), each evolving in (approximately) the familiar classical way. And if this is right, then Everett's notion of a quasi-classical state is itself very different from Gell-Mann and Hartle's (in so far as I understand their notion of a quasi-classical state). Finally, it is worth noting that Everett has already argued that whether or not a system exhibits quantum-mechanical indeterminacy is *independent* of its size (see e.g. Sect. 3.5 above and Everett 1973: 61).

experience? and (2) why should one think that *almost* determinate records are sufficient to explain our *fully* determinate experience?

3.8   *What are branches?*

The recurring problem is that it is difficult to see how Everett's account of an observer's determinate experience is supposed to go even in the simplest situations. Let's give it one more try.

In his thesis Everett says something in his discussion of the qualitative appearance of quantum jumps that provides a clue to how the relativity of states is supposed to work. After pointing out that each term in the final state, when written in the observer's definite memory basis, describes the observer with a definite memory sequence, Everett concludes that '*Relative to him* the (observed) system states are the corresponding eigenfunctions ...' (1957*a*: 24; my italics). One might take this as evidence that he really did believe that there was ultimately no preferred basis for expressing the total state and that there was typically no absolute matter of fact about the state of any subsystem. Indeed, Everett reminds us that

Throughout all of a sequence of observation processes there is only one physical system representing the observer, yet there is no single unique *state* of the observer ... Nevertheless, there is a representation in terms of a *superposition*, each element of which contains a definite observer state and a corresponding system state. (1957*b*: 320, 1957*a*: 25)

But at this point he also begins to talk about the various *branches* of the wave function as if they were somehow real entities. Everett argues that while there is only one physical observer, during a typical measurement the observer state 'splits' into a number of simultaneously existing 'branches', each of which describes the observer as having obtained a determinate, though different, measurement result.

Thus with each succeeding observation (or interaction), the observer state 'branches' into a number of different states. Each branch represents a different outcome of the measurement and the *corresponding* eigenstate for the object-system state. All branches exist simultaneously in the superposition after any given sequence of observations. (1957*b*: 320, 1957*a*: 26)

He adds that

The 'trajectory' of the memory configuration of an observer performing a sequence of measurements is thus not a linear sequence of memory configurations, but a branching tree, with all possible outcomes existing simultaneously in a final superposition with various coefficients in the mathematical model. In any familiar memory device the branching does not continue indefinitely, but must stop at a point limited by the capacity of the memory. (1957*b*: 321, 1957*a*: 26)

And he explains further in a footnote:

Note added in proof.—In reply to a preprint of this article some correspondents have raised the question of the 'transition from possible to actual,' arguing that in 'reality' there is—as our experience testifies—no such splitting of observer states, so that only one branch can ever actually exist. Since this point may occur to other readers the following is offered in explanation.

The whole issue of the transition from 'possible' to 'actual' is taken care of in the theory in a very simple way—there is no such transition, nor is such a transition necessary for the theory to be in accord with our experience. From the viewpoint of the theory *all* elements of a superposition (all 'branches') are 'actual,' none any more 'real' than the rest. It is unnecessary to suppose that all but one are somehow destroyed, since all the separate elements of a superposition individually obey the wave equation with complete indifference to the presence or absence ('actuality' or not) of any other elements. This total lack of effect of one branch on another also implies that no observer will even be aware of any 'splitting' process. (1957*b*: 320–1)

So Everett talks about 'branches' being 'real' and 'splitting' into new 'branches' that are all equally 'actual', and he feels that he needs to explain why this 'splitting' process goes unnoticed. The fact that he uses scare quotes throughout suggests that he is not altogether comfortable with this sort of talk, but there it is, and it has led many readers, in spite of Everett's insistence that the post-measurement states he considers always describe a *single physical observer*, to suppose that he took each term in the superposition to describe a real physical world inhabited by one of the many physical copies of an observer that are created by each observation.

The footnote above continues with an analogy between pure wave mechanics and Copernican astronomy.

Arguments that the world picture presented by this theory is contradicted by experience, because we are unaware of any branching process, are like the criticism of the Copernican theory that the mobility of the earth as a real physical fact is incompatible with the common sense interpretation of nature because we feel no such motion. In both cases the argument fails when it is shown that the theory itself predicts that our experience will be what it in fact is. (1957*b*: 321)

So the explanation is that while it may appear that pure wave mechanics is incompatible with our experience since we do not feel any branching process, just as classical mechanics predicts that one would not feel the earth move, pure wave mechanics predicts that one would not feel any branching process.

Everett claims that it is the total lack of influence of one branch on another that explains why one would not feel the branching process. Since it is unclear what he meant by branches, it is also unclear how this explanation is supposed to go. But regardless of what he meant by branches, if one takes the linear dynamics seriously, then it is generally false that there is a total lack of influence of one branch on another, and Everett knew this.

In his long paper Everett explained that while one might think of a measurement in pure wave mechanics as collapsing the state of the object system into a non-interfering mixture of states just as predicted by the standard formulation of quantum mechanics, the two theories only make the same empirical predictions for measurements made on the original object system. If one were (as in a Wigner's-friend-type story) to measure an observable of the original object system *and the measuring apparatus*, then one must take into account interference effects (1973: 81). After an x-spin measurement, say, pure wave mechanics predicts that the post-measurement state of the measuring device and the object system will be close to

$$\frac{1}{\sqrt{2}}(|x\text{-spin up}\rangle_M|\uparrow_x\rangle_S + |x\text{-spin down}\rangle_M|\downarrow_x\rangle_S), \qquad (3.8)$$

while the standard theory predicts that the final state will be close to either $|x\text{-spin up}\rangle_M|\uparrow_x\rangle_S$ or $|x\text{-spin down}\rangle_M|\downarrow_x\rangle_S$, and there are measurements of the composite system $M + S$ that would, in principle, allow one to determine whether the measurement interaction between M and S led to the pure superposition or a statistical mixture of the two states where the result is determinate. And it is for this reason that pure wave mechanics makes different empirical predictions from the standard collapse formulation of quantum mechanics.

Everett later explained that this is why one cannot suppose that there is only one branch, the one that describes the observer as getting the result that he in fact got, after a measurement.

We take this opportunity to caution against a certain viewpoint which can lead to difficulties. This is the idea that after the apparatus has interacted with the system, in 'actuality' one or the other of the elements of the resultant superposition described by the resultant wave function has been realized to the exclusion of the rest, the existing one simply being unknown to an external observer . . . This position must be erroneous since there is always the possibility for the external observer to make use of interference properties between the elements of the superposition. (1973: 105)

While one can correctly calculate marginal expectations for *subsystems* by supposing that the system is in a mixture of states, 'the representation by a mixture must be regarded as only a mathematical artifice which, although useful in many cases, is an *incomplete description* because it ignores phase relations between the separate elements that actually exist, and which become important in any interactions which involve more than just a subsystem' (1973: 106).[13]

Given that there are experiments that would at least in principle detect the presence of other branches, Everett's explanation why one does not feel a branching process cannot be that there can be no interactions between branches. Rather, a successful explanation would presumably rest on the fact that, where real observers are involved, the sort of interference experiment that would detect the presence of other branches would be extraordinarily difficult to perform. But, as we shall see later (Chapter 8), the difficulty of such measurements is not by itself enough to explain why it seems that we have the determinate measurement records we do.

It is tempting to talk as if we were somehow *in* a particular branch and that this is what explains our particular determinate experience, but Everett himself never says this. It is unclear exactly how Everett thought of branches. It is also unclear whether and in what way a particular observer is associated with a particular branch. What Everett actually says (in his two main passages on the topic) is that there is always a single physical observer, there is no single state of the observer, the different branches represent different subjective experiences of the observer, and all branches exist simultaneously (1957*b*: 320); and he says that since after an observation there are typically many different relative states for an observer, we can speak of different *observers* described by the different states; but since there is only one physical system involved, there is only one observer who has 'different experiences for the separate elements of the superposition' (1973: 68).

---

[13] Such passages seem to me to provide further evidence that decoherence effects had nothing to do with Everett's explanation why observers seem to get determinate measurement results. Indeed, Everett's discussion of statistical mixtures here foreshadows some of the problems with the decoherence theories I shall discuss later (Ch. 8).

Everett says that the various elements of the wave function play the role of assigning states to the *memory* of the observer. 'In conclusion, the continuous evolution of the state function of a composite system in time gives a complete mathematical model for processes that involve an idealized observer. When interaction occurs, the result of the evolution in time is a superposition of states, each element of which assigns a different state to the memory of the observer' (1957*a*: 34). But again, it is difficult to say what Everett meant by assigning many mutually incompatible memories to an observer when we in fact only ever experience one.

Near the end of his long paper Everett describes his new view of physics from the perspective of pure wave mechanics. Physics now consists in the study of quantum-mechanical correlations, and most physical laws are nothing more than statements concerning the quantum-mechanical correlations between subsystems.[14] Such laws can always be written in the form: under conditions *C* the property *A* of a subsystem of the universe is correlated in such-and-such a way with the property *B* of another subsystem (1973: 117–18). But these physical properties do not seem to be determinate properties. Indeed, Everett concludes that there are no absolute physical facts about subsystems: 'All statements about subsystems then become *relative* statements, i.e., statements about the subsystem relative to a prescribed state for the remainder (since this is generally the only way a subsystem even possesses a unique state)' (1973: 118).[15] This again suggests that the particular determinate experience of an observer, like a physical property of a subsystem, is a *relative fact*. But relative to what?

3.9   *Interpreting Everett*

On the title-page of his thesis Everett reported that 'An earlier less condensed draft of the present work, dated January 1956, was circulated to several physicists. Their comments were helpful in the most difficult task of finding the right words to attach to the individual constructs of the present rather straightforward mathematical machinery.' But he laments that 'It would be too much to hope that the revised wording avoids every misunderstanding or ambiguity' (1957*a*: 1).

He was right. The fact that most no-collapse theories have at one time or another been attributed to Everett shows how much the no-collapse

---

[14] Compare this to David Mermin's position (Sect. 7.5).

[15] Such passages suggest something like the relative-fact formulation (Sect. 7.4).

tradition owes to him, but it also shows how hard it is to say what he actually had in mind.

I shall not argue for a particular way of interpreting Everett since I do not believe that there is enough textual evidence to determine exactly what he wanted. I shall not systematically evaluate all of the ways that people have interpreted Everett either. Rather, in the next few chapters, I shall describe a few general approaches for trying to make sense of the determinate experience of observers in the context of Everett's theory. And I shall start with one of the craziest.

4

THE BARE THEORY AND
DETERMINATE EXPERIENCE

I F one accepts Everett's model of a good measuring device and if one insists that the usual deterministic linear dynamics always correctly describes the time-evolution of the quantum-mechanical state, then, as we have seen, an ideal observer M who begins in an eigenstate of being ready to measure the x-spin of a system S that is initially in an eigenstate of z-spin will end up in a post-measurement state like

$$|\psi\rangle = \frac{1}{\sqrt{2}}(|x\text{-spin up}\rangle_M|\uparrow_x\rangle_S + |x\text{-spin down}\rangle_M|\downarrow_x\rangle_S). \quad (4.1)$$

There are several strategies for interpreting $|\psi\rangle$.

One strategy would be to insist that, contrary to what the standard eigenvalue–eigenstate link tells us about $|\psi\rangle$, there is a single post-measurement observer who has recorded a single determinate measurement result. If this is right, then $|\psi\rangle$ must be an *incomplete* description of the state of the composite system $M + S$ since it clearly fails to tell us what result M recorded. Further, even if the coefficients on the two terms in the post-measurement superposition (in the determinate-record basis) were different, one would not want to say that the quantum-mechanical state was complete since we know from experience that any term in the post-measurement state with a nonzero coefficient represents a possible measurement record (and the usual quantum-mechanical state would not tell us which term represented the observer's actual record). On this strategy, one would have to supplement the usual quantum-mechanical state description with a parameter that effectively selects one of the terms in the final state (in the determinate-record basis) as the one that correctly describes what result the observer in fact recorded. That is, if one insists that the usual linear dynamics always correctly describes the time-evolution of the quantum-mechanical state and that there is a single post-measurement observer who records a single determinate measurement result, then one is naturally led to abandon the assumption that the quantum-mechanical state provides a complete description of the post-measurement state of the observer and his object system. Given this, one might try to complete the state by (1) choosing a particular physical quantity as always determinate (the path taken by standard hidden-variable theories) or (2) choosing a rule that itself chooses a determinate physical quantity given the current quantum-mechanical state and the system in which one is interested (the path taken by some so-called modal theories). But in either case, one must also have a rule for determining the *value* of the determinate physical quantity. What quantity is determinate, its value, and the quantum-mechanical state would then together provide a *complete* physical description of a system at a time. I shall discuss such theories in more detail later.

Another strategy for interpreting $|\psi\rangle$ would be to take it as describing a situation where M somehow determinately got *both* x-spin results. One might give up the assumption that there is only one post-measurement physical observer (as one does in a many-worlds theory). Or one might take there to be one physical observer with many mutually independent mental states (as one does in a many-minds theory). Or one might insist that there is only one world and that each observer has only one mind but that one needs a new indexical (akin to time) for discussing facts about an observer *at different branches* akin to how time allows one to discuss facts about an observer *at different times*—one might then say that relative to one branch, the observer records x-spin up, while relative to another branch, the observer records x-spin down, etc. (as one does in a relative-fact theory). But, whichever path one takes, one must explain why it *seems* to an observer that there is a simple matter of fact concerning which measurement result he recorded. The purported advantage of this general strategy is that, unlike the hidden-variable strategy above, the usual quantum-mechanical state can still be understood as being in some sense complete—we do not need to add a parameter that selects a single term in the post-measurement state as representing the observer's actual record since every record represented in the post-measurement state is in some sense equally actual. Most people think that Everett had a theory like this in mind (indeed, most people think he had a many-worlds theory in mind). I shall discuss such theories in more detail later.

Finally, and this is the really crazy one (even crazier, I think, than the relative-fact theory), one might take $|\psi\rangle$ to describe a situation where M recorded no determinate x-spin result at all. To take this strategy seriously, one would have to allow that observers in fact typically (indeed, almost always) fail to record the sort of ordinary determinate measurement results they believe they do. That is, rather than explain why we have the determinate experiences that we have, such a theory would deny that we typically have the sorts of determinate experience we believe we have. Consequently, if such a theory can provide an account of our experience at all, it must be radically different from the sort of account we are used to.

4.1   *The bare theory*

In this chapter I shall consider something that David Albert calls the bare theory. The bare theory is simply the standard von Neumann–Dirac formulation of quantum mechanics with the standard interpretation of states (the eigenvalue–eigenstate link) but stripped of the collapse postulate—hence, *bare*. The bare theory tells us that the global state ψ provides a complete and accurate description of the state of the observer and object system after the measurement. And, given the standard interpretation of states, this means that a typical observer will not determinately get any ordinary measurement result, no matter how insistent he might be. While the bare theory might at first look like a very bad idea, Albert argues that it is in fact 'an amazingly cool idea', and he says that '*this* is the idea that strikes me as interesting to read into Everett's paper' (1992: 124).

While I believe (at least 99.4 per cent of the time) that the bare theory ultimately fails to provide a satisfactory account of our experience, there are none the less several reasons for taking it seriously. Because of the bare theory's formal simplicity, whatever formal properties it has will in one way or another show up in all no-collapse formulations of quantum mechanics. Its simplicity also makes the bare theory a good starting-point for developing a satisfactory no-collapse theory—by carefully considering the problems it faces, one might determine what must be added or changed to get a satisfactory no-collapse theory. There are also good philosophical reasons for studying the ways of the bare theory. The bare theory's account of experience, for example, makes Descartes's demon and other brain-in-the-vat stories look like wildly optimistic appraisals of our epistemic situation. Further, the bare theory raises basic questions concerning the nature of empirical adequacy and even the possibility of empirical inquiry. And finally, the bare theory does indeed turn out to be an interesting way to read Everett. It makes sense of the way Everett sets up the basic problem that his pure wave mechanics must solve: to deduce the subjective appearances of an observer who is in an entangled post-measurement state that does not describe him as having recorded any determinate result. It also allows us to make sense of each of the thought experiments that Everett describes, and it explains how they are relevant to the deduction of subjective appearances—not the same subjective appearances predicted by the standard theory, but appearances that are subjectively indistinguishable from the generic *sorts* of appearance predicted by the standard theory.

4.2   *The suggestive properties*

The bare theory has several suggestive properties.[1] These properties tell us what an observer would report concerning his experience in various measurement situations if the theory were true. While these properties hold for any physical observable, I shall keep things simple by considering only spin observables of spin-$\frac{1}{2}$ systems. I shall describe some of these properties in this section, but save most of the discussion of their significance in interpreting Everett for later sections.

Suppose that M is an x-spin observer in Everett's sense: M can be modelled as an automaton with physical memory registers that represent M's memories of his x-spin measurements. This model requires a close correspondence between physical memory configurations and mental states: if an appropriate memory register is in an eigenstate of recording a particular x-spin result, then the observer M will determinately believe that he got that result. But Everett's commitment to state completeness also suggests that if M determinately believes that he got a particular x-spin result, then the appropriate memory register must be in an eigenstate of recording that result (otherwise, one might argue, the usual quantum-mechanical state would not tell us what the observer's determinate belief was when he determinately had one and would thus be incomplete). I shall suppose, then, that a memory register is in an eigenstate of recording a particular x-spin result if and only if M determinately believes that he got that result. And with respect to what M reports about his own mental state, I shall suppose that if he determinately has the disposition to report that he believes X, then he determinately believes X. This means that one will be able to figure out what M believes either by considering the physical state of his memory registers or by considering his sure-fire dispositions to make various reports about his own mental state.

---

[1]   Some of these properties were first suggested by Everett (1957*a,b*). There have been several subsequent attempts to clarify the properties and to determine their significance: see Hartle (1968), DeWitt (1971), Everett (1973), Graham (1973), Albert and Loewer (1988), Albert (1992), Barrett (1995*a*, 1998) for examples.

I shall also suppose, as I have supposed for observers all along, that M is a *good* x-spin observer in Everett's sense (indeed, one might call him a *perfect* observer): I shall suppose that when a measurement is made, the appropriate memory register becomes perfectly correlated with the x-spin of the S without disturbing it. That is, I shall suppose that M has the two sure-fire dispositions that were discussed earlier: (1) if the initial state is one where M is ready to make a measurement and S is in an x-spin up eigenstate $|r\rangle_M|\uparrow\rangle_S$, then M will evolve to an eigenstate of recording the result *up* and S will still be determinately x-spin up $|\uparrow\rangle_M|\uparrow\rangle_S$ and (2) if the initial state is one where M is ready to make a measurement and S is in an x-spin down eigenstate $|r\rangle_M|\downarrow\rangle_S$, then M will evolve to an eigenstate of recording the result *down* and S will still be determinately x-spin down $|\downarrow\rangle_M|\downarrow\rangle_S$. These two dispositions tell us how M is wired to record his measurement results.

Following Everett, I shall also suppose that M can remember an arbitrarily large number of measurement results and calculate any of their statistical properties. This is clearly an unrealistic idealization, but so are M's other properties.

And now we are ready to reconsider the sequence of thought experiments that Everett described and prove a few simple theorems about the bare theory.

It follows from how M has been wired (from the two dispositions that he has as a good x-spin observer) and from the linear dynamics that if the initial state of $M + S$ is

$$|r\rangle_M(\alpha|\uparrow\rangle_S + \beta|\downarrow\rangle_S), \qquad (4.2)$$

then the state after M's x-spin measurement will be

$$|\psi\rangle = \alpha|\uparrow\rangle_M|\uparrow\rangle_S + \beta|\downarrow\rangle_M|\downarrow\rangle_S. \qquad (4.3)$$

Here M's state has become entangled with S's state. Furthermore, assuming that α and β are both nonzero, $|\psi\rangle$ is not an eigenstate of M reporting a particular determinate x-spin result; rather, it describes M as being part of a superposition of states where it would report mutually incompatible results. That is, $|\psi\rangle$ is not a state where M would report x-spin up and it is not a state where M would report x-spin down. It is, however, a state where M would (falsely) report that it got *some* determinate x-spin result, either x-spin up or x-spin down. This report is false because M in fact fails to have either determinate record.[2]

---

[2] M's state here is sometimes referred to as an *improper* mixture of the two possible record states. According to the eigenvalue–eigenstate link, an improper mixture is not a state

Determinate result. Suppose that M is wired so that it has the disposition to answer the question 'Did you get some determinate result to your x-spin measurement, either x-spin up or x-spin down?' with 'Yes' if it recorded x-spin up (if $M + S$ ended up in the state $|\uparrow\rangle_M|\uparrow\rangle_S$) and with 'Yes' if it recorded x-spin down (if $M + S$ ended up in the state $|\downarrow\rangle_M|\downarrow\rangle_S$). If M in fact recorded a superposition of the two possible x-spin results (if $M + S$ were in the state described by (4.3) above), then it follows from the linearity of the dynamics that he would answer the question 'Did you get some determinate result to your x-spin measurement, either x-spin up or x-spin down?' with 'Yes'; that is, M would report that he got a determinate x-spin result when he did not determinately get up and did not determinately get down.

Asking M whether he got a determinate x-spin result here amounts to measuring a physical observable of $M + S$. Let $\hat{D}$ be an observable such that eigenvalue $+1$ corresponds to a state where M has the disposition to report 'I did get a determinate result to my x-spin measurement' and eigenvalue -1 corresponds to any orthogonal states. Since $|\uparrow\rangle_M|\uparrow\rangle_S$ corresponds to a state where M has recorded $\uparrow$ for the outcome of its measurement, if M is operating correctly (that is if M is operating according to the dispositions described earlier), he will report that it obtained a determinate x-spin result when in this state; and since the same is true for $|\downarrow\rangle_M|\downarrow\rangle_S$, both of these are eigenstates of $\hat{D}$ with eigenvalue $+1$. That is,

$$\hat{D}|\uparrow\rangle_M|\uparrow\rangle_S = |\uparrow\rangle_M|\uparrow\rangle_S \text{ and } \hat{D}|\downarrow\rangle_M|\downarrow\rangle_S = |\downarrow\rangle_M|\downarrow\rangle_S. \qquad (4.4)$$

So

$$\begin{aligned}
\hat{D}|\psi\rangle &= \hat{D}(\alpha|\uparrow\rangle_M|\uparrow\rangle_S + \beta|\downarrow\rangle_M|\downarrow\rangle_S) \\
&= \alpha\hat{D}|\uparrow\rangle_M|\uparrow\rangle_S + \beta\hat{D}|\downarrow\rangle_M|\downarrow\rangle_S \\
&= \alpha|\uparrow\rangle_M|\uparrow\rangle_S + \beta|\downarrow\rangle_M|\downarrow\rangle_S \\
&= |\psi\rangle.
\end{aligned} \qquad (4.5)$$

Consequently, $|\psi\rangle$ is an eigenvector of $\hat{D}$ with eigenvalue $+1$, which is just another way of saying that M has the sure-fire disposition to report 'I got a determinate result to my x-spin measurement, either x-spin up or x-spin down.'

where M has a determinate record. It is important to distinguish this state from a *proper* mixture; a state in which M has in fact recorded one or the other of the two possible results but we do not know which.

Consider what this means. Suppose that the bare theory is true, and that an observer begins a measurement in an initial state like the one described above (4.1) (a remarkably unlikely event if the bare theory is true, but let's set this aside for now). If the observer is competent at reporting his beliefs when he determinately observes a pointer in an eigenstate of pointing at x-spin up and when he determinately observes a pointer in an eigenstate of pointing at x-spin down, then by the argument above, when he ends up in a superposition of believing that he sees x-spin up and believing that he sees x-spin down, he will report that he got a determinate result; that is he will answer the question 'Did you get a determinate result of either x-spin up or x-spin down?' with 'Yes', and this is what he would actually believe (assuming that the observer believes what he has the sure-fire disposition to report). Further, he would believe that he knows what that result is (because this is precisely what he would have the sure-fire disposition to report in each situation where he did in fact record a determinate result). But both of these beliefs would be false because he did not in fact determinately record either of the two results (and while he may claim to know *which* determinate result he got, there is in fact no specific determinate result that he believes that he got). A proponent of the bare theory would believe that a significant portion of the experience of real observers can be explained as this sort of illusion: a situation where an observer (falsely) believes that he has an ordinary determinate experience.

One might object that the operator $\hat{D}$ that is supposed to correspond to the question 'Did you get some determinate result to your x-spin measurement?' is the identity here and that such a simple operator cannot possibly represent such an interesting question (for just such an objection, see Weinstein 1996). First, it is probably worth noting that $\hat{D}$ is not really the identity: while every linear superposition of $|\uparrow\rangle_M|\uparrow\rangle_S$ and $|\downarrow\rangle_M|\downarrow\rangle_S$ are eigenstates of $\hat{D}$ with eigenvalue $+1$, $|r\rangle_M|\uparrow\rangle_S$, for example, is not $(\hat{D}|r\rangle_M|\uparrow\rangle_S = -|r\rangle_M|\uparrow\rangle_S)$. But the important point here is that how one represents the observable corresponding to asking M whether he got a determinate result is irrelevant. What matters is that if M were constructed with the dispositions described above, if he were wired to report that he recorded a determinate result in those situations where he did in fact record a determinate result (which is how we believe that we are wired), then the usual linear dynamics entails that he would have the *sure-fire disposition* to report that he got a determinate x-spin result when he was actually in an entangled superposition of recording mutually incompatible x-spin results.

Repeatability. Suppose M makes a second x-spin measurement. Let M_1 and M_2 be the registers M uses to record the first and second measurement results. If S is undisturbed between measurements, the state $|\psi_2\rangle$ after the second measurement will be

$$\alpha|\uparrow\rangle_{M_1}|\uparrow\rangle_{M_2}|\uparrow\rangle_S + \beta|\downarrow\rangle_{M_1}|\downarrow\rangle_{M_2}|\downarrow\rangle_S. \qquad (4.6)$$

Let $\hat{C}$ be an observable such that states where M has the disposition to report that his first and second x-spin measurements agree (that is, they are both $\uparrow$ or both $\downarrow$) correspond to eigenvalue $+1$. Since registers M_1 and M_2 agree in the states represented by $|\uparrow\rangle_{M_1}|\uparrow\rangle_{M_2}|\uparrow\rangle_S$ and $|\downarrow\rangle_{M_1}|\downarrow\rangle_{M_2}|\downarrow\rangle_S$, it follows that

$$\hat{C}|\uparrow\rangle_{M_1}|\uparrow\rangle_{M_2}|\uparrow\rangle_S = |\uparrow\rangle_{M_1}|\uparrow\rangle_{M_2}|\uparrow\rangle_S$$
$$\text{and} \quad \hat{C}|\downarrow\rangle_{M_1}|\downarrow\rangle_{M_2}|\downarrow\rangle_S = |\downarrow\rangle_{M_1}|\downarrow\rangle_{M_2}|\downarrow\rangle_S. \qquad (4.7)$$

Since $|\psi_2\rangle$ is just a linear combination of eigenvectors of $\hat{C}$ with eigenvalue $+1$, it is also an eigenvector of $\hat{C}$ with eigenvalue $+1$, so M has the disposition to report that his first and second x-spin measurements agree.

Agreement. Suppose that rather than being interpreted as registers of the same observer, M_1 and M_2 are interpreted as different observers capable of comparing their results. It immediately follows from the repeatability property above that if M_1 and M_2 compare their results, they will have the sure-fire dispositions to report that their measurement results agree (even though neither in fact recorded a determinate result).

These three properties tell us that if M makes reliable reports about his beliefs when he does end up with determinate records, then if M starts in an eigenstate of being ready to measure the x-spin of S, he will report that he got a determinate x-spin result, *up* or *down*, even if S is initially in a superposition of x-spin eigenstates; if M carefully remeasures S, then he will report that his second result agrees with his first; and if a second perfect observer measures the x-spin S and the two observers compare their results, then both will report that their results agree. One might conclude, then, that it would *seem* to each observer that the state of the object system had collapsed to exactly one of the eigenstates of the observable being measured when the time-evolution of the composite system was in fact perfectly deterministic and continuous. This, of course, is part of what Everett wanted—there is a sense in which one might say that the bare theory allows one to deduce the phenomena of a collapse as the subjective appearances of an observer treated within the context of the theory. But again, if α and β are both nonzero, then according to the standard interpretation of states, an observation will in fact typically not yield a determinate measurement record, at least not in the ordinary sense, which is presumably what one would need to deduce the *same* subjective experiences predicted by the standard theory. I shall consider the sense in which the bare theory does and does not predict the same appearances as the standard theory in more detail after I discuss the bare theory's statistical properties.

Relative frequency. Consider a system T consisting of an observer M and an infinite set of systems $S_1, S_2, S_3, \ldots, S_n, \ldots$, each of which is initially in the state $\alpha|\uparrow\rangle_{S_n} + \beta|\downarrow\rangle_{S_n}$, where $|\uparrow\rangle_{S_n}$ and $|\downarrow\rangle_{S_n}$ are x-spin eigenstates and α and β are nonzero. Let T_n be the system consisting of M's first n registers and systems S_1 through S_n. The Hilbert space where one represents the state of T_n is

$$\mathcal{H}_n = \mathcal{M} \otimes \mathcal{S}_1 \otimes \cdots \otimes \mathcal{S}_{n-1} \otimes \mathcal{S}_n, \qquad (4.8)$$

where $\mathcal{M}$ is the Hilbert space representing the first n registers of the measuring device M, $\mathcal{S}_1$ is the Hilbert space representing the system S_1, etc.

Suppose M makes an x-spin measurement on each S_n in turn. The states before and after the nth measurement might be represented by elements of $\mathcal{H}_n$. For example, the state of T_1 before the first measurement is represented by the vector

$$|r\rangle_M(\alpha|\uparrow\rangle_{S_1} + \beta|\downarrow\rangle_{S_1}), \qquad (4.9)$$

which is an element of $\mathcal{H}_1$. After the first measurement, T_1 will be in the state represented by

$$\alpha|\uparrow\rangle_M|\uparrow\rangle_{S_1} + \beta|\downarrow\rangle_M|\downarrow\rangle_{S_1}. \qquad (4.10)$$

Similarly, the state of T_2 in $\mathcal{H}_2$ before the second measurement is the state represented by

$$(\alpha|\uparrow, r\rangle_M|\uparrow\rangle_{S_1} + \beta|\downarrow, r\rangle_M|\downarrow\rangle_{S_1})(\alpha|\uparrow\rangle_{S_2} + \beta|\downarrow\rangle_{S_2}), \qquad (4.11)$$

and after the second measurement,

$$\alpha^2 |\uparrow, \uparrow\rangle_M |\uparrow\rangle_{S_1} |\uparrow\rangle_{S_2} + \alpha\beta |\uparrow, \downarrow\rangle_M |\uparrow\rangle_{S_1} |\downarrow\rangle_{S_2}$$
$$+ \beta\alpha |\downarrow, \uparrow\rangle_M |\downarrow\rangle_{S_1} |\uparrow\rangle_{S_2}$$
$$+ \beta^2 |\downarrow, \downarrow\rangle_M |\downarrow\rangle_{S_1} |\downarrow\rangle_{S_2}. \tag{4.12}$$

There are 2^n terms in the vector $|\psi_n\rangle$ representing the state of T_n after n measurements when written in the determinate-record basis.

Choose a determinate-record basis for M in each space $\mathcal{H}_n$, a basis such that each element describes a determinate memory configuration for the observer in the states described in this experiment. Call this the $\hat{X}_n$-basis, and let the observable $\hat{X}_n$ be such that T_n is in an eigenstate of $\hat{X}_n$ if and only if M is determinately reporting a particular sequence of results for the first n measurements. Given this, one can define a family of relative-frequency operators $\hat{F}_n(\epsilon)$ such that $\hat{F}_n(\epsilon)|\psi_n\rangle = |\psi_n\rangle$ if and only if $\hat{X}_n|\psi_n\rangle = \lambda|\psi_n\rangle$ and $(n-m)/n = |\alpha|^2 \pm \epsilon$, for $\epsilon > 0$, where $n - m$ is the number of $\uparrow$-results that M has after n measurements. That is, if T_n is in an eigenstate of $\hat{F}_n(\epsilon)$, then M is in an eigenstate of reporting that the ratio of the number of $\uparrow$-results $n - m$ to the total number of results n is within ϵ of $|\alpha|^2$—that is, that the ratio of $\uparrow$-results to $\downarrow$-results is about what the standard theory would predict. Note, however, that since $|\psi_n\rangle$ is not an eigenvector of $\hat{X}_n$ for any finite n, it is also not an eigenvector of $\hat{F}_n(\epsilon)$ for any finite n.

Write $|\psi_n\rangle$ in the $\hat{X}_n$-basis, and let $|\chi(\epsilon)_n\rangle$ be the sum of those terms where the ratio of the number of $\uparrow$-results $n - m$ to the total number of results n is within ϵ of $|\alpha|^2$. In other words, let $|\chi(\epsilon)_n\rangle$ be the sum of the terms in the $\hat{X}_n$-expansion of $|\psi_n\rangle$ that *are* eigenvectors of $\hat{F}_n(\epsilon)$ with eigenvalue $+1$. That is, $|\chi(\epsilon)_n\rangle$ is the sum of the terms in the determinate-record expansion of the state after n measurements where the relative frequencies are about what the standard collapse theory would predict. Note that since $|\langle\chi(\epsilon)_n|\chi(\epsilon)_n\rangle|^2 < 1$ for all finite n, $|\chi(\epsilon)_n\rangle$ does not correspond to the state of any system; rather, $|\chi(\epsilon)_n\rangle$ is just the orthogonal projection of the state $|\psi_n\rangle$ onto the $\lambda = +1$ eigenspace of $\hat{F}_n(\epsilon)$. Finally, note that for all n, $|\chi(\epsilon)_n\rangle$ is an eigenvector of $\hat{F}_n(\epsilon)$ with eigenvalue $+1$.

The following lemma says that, for all $\epsilon > 0$, the magnitude of the component of $|\psi_n\rangle$ that is an eigenvector of $\hat{F}_n(\epsilon)$ with eigenvalue $+1$ (that is, the magnitude of $|\chi(\epsilon)_n\rangle$) goes to one as n gets large. We will then interpret this to mean that T_n approaches a physical state where the observer M has the sure-fire disposition of reporting the usual quantum statistics for his measurements.

Lemma *For all $\epsilon > 0$, $\lim_{n \to \infty} |\langle \chi(\epsilon)_n | \chi(\epsilon)_n \rangle|^2 = 1$.*

Proof. Written in the $\hat{X}_n$-basis, the vector representing the state of T_n after n measurements $|\psi_n\rangle$ is

$$
\begin{aligned}
\alpha^n |\uparrow, \uparrow, \ldots, \uparrow\rangle_M &|\uparrow\rangle_{S_1} |\uparrow\rangle_{S_2} \ldots |\uparrow\rangle_{S_n} \\
&+ \alpha^{n-1}\beta |\uparrow, \uparrow, \ldots, \downarrow\rangle_M |\uparrow\rangle_{S_1} |\uparrow\rangle_{S_2} \ldots |\downarrow\rangle_{S_n} \\
&+ \alpha^{n-2}\beta^2 |\uparrow, \uparrow, \ldots, \downarrow, \downarrow\rangle_M |\uparrow\rangle_{S_1} |\uparrow\rangle_{S_2} \ldots |\downarrow\rangle_{S_{n-1}} |\downarrow\rangle_{S_n} \\
&+ \ldots \\
&+ \beta^n |\downarrow, \downarrow, \ldots, \downarrow\rangle_M |\downarrow\rangle_{S_1} |\downarrow\rangle_{S_2} \ldots |\downarrow\rangle_{S_n}
\end{aligned}
\tag{4.13}
$$

with 2^n such terms.

I shall first partition these terms into equivalence classes by the relative frequency of $\uparrow$-results in each term, then consider a measure μ_n that assigns a real number to each equivalence class. For a given n, all terms where the number of $\uparrow$-results equals $n-m$ will be in the same equivalence class $\mathcal{B}_m$. For a given n, $\mu_n(\mathcal{B}_m)$ will be the sum of the squares of each of the coefficients of the terms in $\mathcal{B}_m$. For a given n, then, there will be n equivalence classes, and

$$
\mu_n(\mathcal{B}_m) = \binom{n}{m} |\alpha^{n-m}\beta^m|^2,
\tag{4.14}
$$

which means that

$$
\sum_m \binom{n}{m} |\alpha^{n-m}\beta^m|^2 = |\langle \chi(\epsilon)_n | \chi(\epsilon)_n \rangle|^2,
\tag{4.15}
$$

where the sum is over all m such that $m \leq n$ and $|\alpha|^2 - \epsilon \leq (n-m)/n \leq |\alpha|^2 + \epsilon$—that is, the sum is over all m where the ratio of the number of $\uparrow$ results to the total number of results is within ϵ of $|\alpha|^2$. Since it is a basic result of probability theory that

$$
\lim_{n \to \infty} \sum_m \binom{n}{m} |\alpha^{n-m}\beta^m|^2 = 1,
\tag{4.16}
$$

$|\langle \chi(\epsilon)_n | \chi(\epsilon)_n \rangle|^2 \to 1$ as $n \to \infty$. $\qquad \square$

Given this, it is easy to show that, as n gets large, $|\psi_n\rangle$ gets arbitrarily close to an eigenvector of $\hat{F}_n(\epsilon)$ with eigenvalue $+1$ (that is, the state of the composite system gets closer and closer to a state where M would have the sure-fire disposition to report that his results exhibit the standard relative frequencies).

Theorem *For all $\epsilon > 0$, $\lim_{n\to\infty} |\langle \psi_n - \chi(\epsilon)_n | \psi_n - \chi(\epsilon)_n \rangle|^2 = 0$.*

Proof. Since $|\chi(\epsilon)_n\rangle$ is an orthogonal projection of $|\psi_n\rangle$ onto a subspace of $\mathcal{H}_n$, namely the eigenspace corresponding to eigenvalue $+1$ of $\hat{F}_n(\epsilon)$, $|\psi_n - \chi(\epsilon)_n\rangle$ is orthogonal to $|\chi(\epsilon)_n\rangle$. So, by the Pythagorean theorem,

$$|\langle \psi_n - \chi(\epsilon)_n | \psi_n - \chi(\epsilon)_n \rangle|^2 = |\langle \psi_n | \psi_n \rangle|^2 - |\langle \chi(\epsilon)_n | \chi(\epsilon)_n \rangle|^2.$$
(4.17)

Since $|\langle \psi_n | \psi_n \rangle|^2 = 1$ for all n and since $|\langle \chi(\epsilon)_n | \chi(\epsilon)_n \rangle|^2 \to 1$ as $n \to \infty$, $|\langle \psi_n - \chi(\epsilon)_n | \psi_n - \chi(\epsilon)_n \rangle|^2 \to 0$ as $n \to \infty$. $\square$

It would be nice if one could use von Neumann's formalism to conclude that the composite system converges to an *eigenstate* of the appropriate relative-frequency operator in the limit, but one cannot conclude this here because the space in the limit $\mathcal{H}_\infty$ is nonseparable and so it does not even qualify as a Hilbert space on von Neumann's definition (see condition E in Appendix A). Consequently, I shall simply interpret the fact that the state of the composite system gets arbitrarily close to a state where the observer M would report that he got the standard relative frequencies for his first n measurements as n gets large to mean that the composite system approaches a physical state where M would report that he got the usual relative frequencies.[3]

So, given this assumption about limiting states, it follows from the above theorem that if M makes measurements of the x-spin on each system, then the composite system will approach a state where M would answer the question 'Were your results distributed with the usual quantum

---

[3] If one does not like this way of talking, then one can read 'approaches a state where ... ' as 'for any distance in the appropriate Hilbert-space metric, there are a number of measurements n that put the composite system within the specified distance of a state where M would report that his results after n measurements were distributed with the usual quantum relative frequencies'. But there is little reason for such squeamishness since there are richer mathematical formalisms than von Neumann's for representing quantum-mechanical states where such processes have perfectly well-defined limits. See e.g. Farhi *et al.* (1989). The important thing to note here is that as the difference in physical states decreases, so does the difference in physical dispositions. Along these lines it is also worth noting that this experiment can be reformulated as a two-box experiment like that discussed at the end of Ch. 5. (M would not keep track of his results but only the current relative frequency of *ups*, say. The yes–no question would be 'Is the relative frequency within ϵ of the quantum relative frequency?' In the limit, and there *is* a Hilbert-space limit here, the wave function would zero outside the *yes* box.)

relative frequencies?' with 'Yes' as the number of observations gets large. Note that while the composite system approaches a state where M would report that he has a determinate sequence of measurement records that are distributed with precisely the same relative frequencies predicted by ordinary quantum mechanics, if α and β are nonzero, then M in fact has no sequence of determinate measurement records. Further, note that M will *always* approach a state where he makes this report, not *almost always* as the collapse dynamics would predict. Thus there is a subtle difference in the nature of the limiting statistics in the bare theory and the standard collapse theory.

Randomness. As the number of x-spin measurements gets large, the composite system also approaches a state where M would answer the question 'Were your results randomly distributed?' with 'Yes'.

Proof. Given the argument for the relative-frequency property, this is fairly straightforward. Write $|\psi_n\rangle$, the state of T_n after n measurements, in the $\hat{X}_n$-basis (the determinate-record basis). Each term in the expansion will correspond to a different sequence of measurement results, and each length-n sequence will be represented by some term in the expansion. Let A_n be the set of *all possible* length-n sequences of measurement results, and let $R_n \subset A_n$ be the set of *random* length-n sequences (on any standard notion of random).[4] Let $\mu(R_n)$ be the sum of the norm-squared coefficients on each term of $|\psi_n\rangle$ that describes a random sequence of results. If $\mu(R_n)$ goes to one as n gets large, then one can conclude, as in the case of the relative-frequency property, that the composite system approaches a state where M would report that his measurement results were randomly distributed. Since every possible length-n sequence of results is represented by exactly one term in the determinate-record expansion of $|\psi_n\rangle$, since the norm-squared of the coefficient of each term goes to zero as n gets large, and since the proportion of length-n sequences that are random goes to one as n gets large (on any of the standard notions of what it means to be a random sequence), $\mu(R_n)$ goes to one as n gets large. So the composite system approaches a state where M would report that his measurement results are randomly distributed (on any standard notion of a random sequence, and more generally on any notion of a random

---

[4] One might, for example, choose an algorithmic language $\mathcal{L}$, then say that a finite sequence of results is *random* if and only if the shortest algorithm (written in $\mathcal{L}$) that prints the sequence and stops is longer than the string itself (and one might say that an infinite sequence is random if and only if there is no finite-length algorithm that prints that sequence). See Chaitin (1987).

sequence where the proportion of random length-n sequences goes to one in the limit).

Combining the relative-frequency and randomness properties, an observer who repeated the same x-spin measurement on identically prepared systems would approach a state where he would report that his results were randomly distributed with the same relative frequencies predicted by the standard collapse theory. But, of course, there is nothing special about x-spin measurements: if *any experiment* yields a state where the amplitude of M recording the result r is α, then the composite system will approach a state where M will report that the result r is randomly distributed with the usual quantum-mechanical relative frequency $|\alpha|^2$ in the limit as the number of identical experiments gets large. I shall call this the *general limiting property*. The general limiting property explains why the composite system would approach a state where an observer would report the usual quantum statistics for any experiment as that experiment is repeated an infinite number of times—even if each experiment involves making a series of measurements of possibly incompatible quantum observables.[5]

Perhaps an example would help to clarify how the general limiting property works in predicting beliefs about joint probabilities. Suppose that two observers A and B are prepared to make space-like separate x-spin measurements of systems S_A and S_B respectively and that the two systems were initially in the EPR state

$$\frac{1}{\sqrt{2}}(|\downarrow_x\rangle_{S_A}|\uparrow_x\rangle_{S_B} - |\uparrow_x\rangle_{S_A}|\downarrow_x\rangle_{S_B}). \tag{4.18}$$

When A and B perform their measurements, they become entangled with S_A and S_B. If one writes the state of the composite system in the A–B pointer basis, then the square of the coefficient on each term gives the standard quantum probability of getting the joint result at A and B that is

---

[5] Instead of considering each experiment to consist of a single x-spin measurement, suppose that each experiment consists of an arbitrary finite sequence of measurements. Note that if the coefficient on the term where the sequence of results is r is α after the first experiment, then as a series of identical experiments are performed, the composite system will approach a state where the observer will report that his results were randomly distributed and that the relative proportion of the sequence r was $|\alpha|^2$ by the same sort of argument as given for x-spin measurements above. Note that this simple argument works for any repeated experiment with a *finite* number of possible results.

described by that term (see Appendix B for a concrete example of what an EPR experiment would look like in the bare theory). It follows from the general limiting property that as this EPR experiment is repeated, the composite system will approach a state where *A* and *B* would report exactly the same statistical correlations predicted by the standard collapse theory. Since the probabilities predicted by the standard theory fail to satisfy the Bell-type inequalities, the bare theory predicts that the composite system will approach a state where *A* and *B* report that their results failed to satisfy the Bell-type inequalities.[6] Observers in the bare theory might consequently be tempted to conclude that there is some sort of nonlocal causal connection between their measurements. But note that since the usual linear dynamics can be expressed in a perfectly local form, there are in fact no nonlocal causal influences here. The bare theory explains the appearance of nonlocality in a way that is perfectly compatible with special relativity.[7] A feature of this account of the EPR statistics is that while the observers would conclude that their measurement results violated the Bell-type inequalities, there would in fact fail to be any determinate measurement records. The bare theory thus provides a rather curious way of understanding Everett's claim that nonlocality is only *apparent* in his theory.

Together with the determinate-result property, the general limiting property entails that an observer will believe that he got a perfectly determinate result to each experiment, and, if he is persistent, then he will approach a state where he believes that his results were randomly distributed with the same relative frequencies (and joint relative frequencies) as predicted by the standard collapse theory (whenever it makes a coherent prediction). This is what Everett wanted in his deduction of the predictions of the standard collapse theory as subjective appearances.[8]

---

[6] The Bell-type inequalities are inequalities between probabilities like the one described by Bell himself in his famous EPR paper. There are now many such inequalities that one might naively expect to hold, but which are violated by the statistical predictions of quantum mechanics (and our best empirical results).

[7] Remember that one problem with the standard formulation of quantum mechanics is that the collapse dynamics fails to be Lorentz-covariant, which makes it incompatible with relativity. Since the usual linear dynamics can be written in a covariant form, in so far as the bare theory can explain the statistics reported by observers, it can do so in a perfectly covariant way. But while the statistical properties reported by an observer in the standard collapse theory are properties of his actual measurement results, the statistical properties reported by an observer in the bare theory are not the statistical properties of any determinate results.

[8] Or perhaps we should say that this is *close* to what Everett wanted. We are, after all, still talking about the *limiting* properties of the composite system. If the measurements are made at equal time-intervals (and if we assume the standard eigenvalue–eigenstate link),

But the observer will in fact typically fail to have any determinate records or beliefs concerning *which* specific sequence of results he recorded, so the statistical properties that he would report in the limit would not be the statistical properties of any *actual* determinate sequence of results. And this is presumably not what Everett wanted.

4.3   *Imperfect measurements*

So far we have assumed that all measurements are perfect—that all measurement interactions induce a *perfect* correlation between the state of an observer's most immediately accessible physical records and the observed quantity of the object system without disturbing the property being measured. In practice, however, neither of these assumptions are typically warranted. Further, if the usual linear dynamics always correctly describes the time-evolution of every physical system, then it is highly unlikely that the state of the world is ever one where an observer determinately performs *any* experiment. I shall continue to make this last assumption, and consider here what happens when one drops the first two.

An imperfect x-spin measurement might look something like the following. Suppose that before the measurement the state of the observer M and his object system S is represented by

$$|r\rangle_M(\alpha|\uparrow\rangle_S + \beta|\downarrow\rangle_S). \qquad (4.19)$$

After an imperfect measurement, rather than

$$\alpha|\uparrow\rangle_M|\uparrow\rangle_S + \beta|\downarrow\rangle_M|\downarrow\rangle_S, \qquad (4.20)$$

then M never actually aquires the sure-fire disposition to report the appropriate quantum statistics. One might try to get closer to what Everett wanted by weakening the eigenvalue–eigenstate link so that M would have the sure-fire disposition to make the appropriate reports when the composite system gets *close enough* (in the Hilbert-space metric) to the appropriate state. As we shall see later, weakening the eigenvalue–eigenstate link in this way can have curious consequences. Another way that one might get closer to what Everett wanted would be to suppose that all the measurements were made within a finite time: the measurements at, say, 1/2 second, 3/4 second, 7/8 second, etc. In so far as such a sequence of measurements is physically possible (see Earman and Norton 1996 for a nice discussion of the physical possibility of such *supertasks*), M would presumably have the appropriate sure-fire dispositions at 1 second. One problem with this second strategy is that since we presumably never make such a sequence of measurements, it is unclear how such a story is supposed to be relevant to our experience.

the composite system will be in a state like

$$\alpha'|\uparrow\rangle_M|\uparrow\rangle_S + \beta'|\downarrow\rangle_M|\downarrow\rangle_S$$
$$+ \gamma_1|\downarrow\rangle_M|\uparrow\rangle_S + \gamma_2|\uparrow\rangle_M|\downarrow\rangle_S$$
$$+ \gamma_3\left[|r\rangle_M(\alpha|\uparrow\rangle_S + \beta|\downarrow\rangle_S)\right]$$
$$+ \gamma_5|\text{register correlated with the wrong}$$
$$\text{spin observable}\rangle_M \cdots$$
$$+ \gamma_6|\text{blown to bits}\rangle_M \cdots$$
$$+ \cdots, \tag{4.21}$$

where $|\alpha'|^2 \approx |\alpha|^2$, $|\beta'|^2 \approx |\beta|^2$, and all of the γ-coefficients are small (in the obvious sense). Clearly, M is not in an eigenstate here of reporting that he got either x-spin up or x-spin down as the result of his measurement. But if the coefficients are zero on all the terms that describe the observer as no longer being capable of making a report (like the term with coefficient γ_6 above), then the definite-result property discussed earlier has a natural extension. After the above imperfect measurement, while the observer is not in an eigenstate of answering the question 'Did you get a definite experimental result of either x-spin up or x-spin down?' with 'Yes', he is in an eigenstate of answering the question 'Did you get a definite experimental result of either x-spin up, x-spin down, measuring-device failure, etc.?' with 'Yes'. The point is that no matter how bad the measurement is, as long as the observer is still determinately able to make a report, there is always a question that one can ask such that his sure-fire response will suggest that the measurement yielded *some* perfectly ordinary determinate result. The determinate-result property then is relatively robust.

The repeatability and agreement properties, however, are very sensitive to measurement imperfections (as we noted in the context of Everett's own thought experiments). A real observer would typically have a non-zero amplitude of disturbing his object system over the course of his measurement, and he would probably not repeat exactly the same errors in every measurement. Consequently, one would not expect an observer to end up in a state where he determinately reports that he got the same result when he repeated his measurement, nor would one expect two observers to end up in a state where they determinately agreed that their results agreed. But the general limiting property is somewhat less sensitive to measurement imperfections.

If all imperfections in an observer's x-spin measurements consist of the measurements being made in directions slightly off the x-axis (and thus correspond to measurements of spin observables slightly different from x-spin) and if these errors are randomly distributed in such a way that there is an average (mean) spin measurement of the observer's actual spin measurements in the limit as the number of measurements gets large, then one can show that an observer will approach a state where he has the sure-fire disposition to answer the question 'What was the relative frequency of spin-up results to the total number of measurements that yielded definite spin results?' with the usual quantum statistics *for the average spin measurement.*[9] This fact might then be used to provide an account of repeatability and agreement in imperfect experiments in terms of reports that observers would make in the limit as they perform many imperfect experiments about how often they found that repeated measurements yielded the same results.

All this is fine as far as it goes, but there is a rather serious problem: every thought experiment we have considered so far depends on an observer and the systems he observes beginning the experiment in a very special state. The observer must be in an eigenstate of being alive, conscious, in the right place with a properly functioning measuring device, etc., and the systems he observes must determinately exist, be in the right location to be observed, etc. But if the bare theory is right, then, given the nature of the usual linear dynamics, one would expect such special conditions virtually never to be met. Further, we have assumed that the observer is in an eigenstate of being sentient after an experiment as well, but, given the linear dynamics, one would not expect this condition to be satisfied in a real experiment even if, by some miracle, the special initial conditions were met. And finally, where the limiting properties are concerned, it is difficult to say what their relevance to our experience *could be*, given that we never in fact perform such infinite sequences of measurements. But let's back up a little and try to see how far the bare theory can go in explaining our experience.

---

[9] Such an experiment is analogous to an infinite coin-toss experiment where the probability of heads changes from toss to toss but there is none the less a mean probability of heads p^* in the limit—one would expect almost all sequences of tosses (in the product measure generated by p^*) to have a relative frequency of heads of $p*$ in the limit. Similarly, almost all of the terms in the determinate-result expansion (in the squared-coefficient measure) will eventually exhibit appropriate relative frequencies in a sequence of imperfect x-spin measurements.

4.4   *The account of experience*

So just how far can the bare theory go in explaining our experience? More specifically, does the bare theory allow one to deduce the empirical predictions made by the standard theory as subjective appearances as Everett wanted? Can we get anything *like* the right empirical predictions?

As we have seen, one can cook up a reasonable-sounding argument that an observer who begins in a state where he is ready to make a series of measurements on various systems would believe (have the appropriate sure-fire dispositions) that he got a determinate result for each measurement and that the composite system will approach a state where the observer would believe that his results were randomly distributed according to the usual quantum statistics as he makes enough measurements of each type of similarly prepared system. This sounds very much like what Everett wanted. But even in a very special story like this (a story where the observer begins in an eigenstate of being ready to perform his measurements, etc.), there is an important sense in which one cannot deduce the standard theory's empirical predictions as subjective appearances. Rather than predict the same determinate appearances as predicted by the standard theory, the bare theory seeks to explain why one might *falsely believe* that one had determinate appearances of the *sort* predicted by the standard theory.

In order to be clear about the way in which the bare theory accounts for our experiences and beliefs, it is useful to distinguish between *ordinary* and *disjunctive* records, experiences, and beliefs. Suppose that an observer M measures the x-spin of a system in an eigenstate of z-spin and thus ends up in a superposition of recording x-spin up and x-spin down. It follows immediately from the standard eigenvalue–eigenstate link (and the assumption that we can talk about an observer having a belief the same way that we talk about a measuring device having a physical record) that in this state M does not believe x-spin up, does not believe x-spin down, does not believe both, and does not believe neither. Thus, a proponent of the bare theory cannot say that either M would determinately believe x-spin up or that M would determinately believe x-spin down after the measurement. And this means that if one insists that exactly one of the two ordinary beliefs is what M would in fact believe, then the bare theory cannot account for M's experience. But again, and this is a critically important point that is easy to miss, a proponent of the bare theory would *deny* that M would end up with either of these two ordinary determinate beliefs—he would not say that M would determinately believe that he had recorded x-spin up, nor would he say that he would believe that he had recorded x-spin down; rather, he would say that M would determinately believe that he had recorded x-spin up *or* x-spin down. One might call the experience leading to this disjunctive belief a disjunctive experience. This disjunctive experience would be phenomenally indistinguishable from either getting x-spin up *or* getting x-spin down *because this is precisely what the observer would have the sure-fire disposition to report* (and we are assuming that he believes what he reports). But it would be wrong to say that it would be phenomenally indistinguishable from getting x-spin up and it would be wrong to say that it would be phenomenally indistinguishable from getting x-spin down since the observer would not be in an eigenstate of making either of these reports. And, for the same reason, it would also be wrong to say that the disjunctive experience would be phenomenally distinguishable from getting x-spin up or that it would be phenomenally distinguishable from getting x-spin down. Again, the right thing to say is that the observer would be unable to distinguish the disjunctive result from x-spin up *or* x-spin down.

The bare theory, then, does not seek to account for the sort of ordinary determinate experiences that we (naively, according to the bare theory) suppose ourselves to have. Rather, the bare theory denies that there typically are any such phenomenal experiences, then seeks to explain why one might none the less believe that there were—why one would mistake *disjunctive* experiences for *generic nondisjunctive* experiences (*generic* because one would find a particular disjunctive experience neither distinguishable nor indistinguishable from a given *specific* associated ordinary experience).

Perhaps the following thought experiment will help to make the distinction between ordinary and disjunctive records, experiences, and beliefs clearer. Suppose that an observer measures the x-spin of three object systems: the first is in an x-spin up eigenstate, the second is in an x-spin down eigenstate, and the third is in a superposition of x-spin up and x-spin down. The observer will believe that he has a determinate phenomenal result in all three cases. Moreover, he will believe that the result of his last measurement was phenomenally indistinguishable from the result of exactly one of his first two measurements. But the observer's disjunctive result will be neither distinguishable nor indistinguishable from getting x-spin up (the observer will not have a determinate belief concerning whether his first and third results agree) and it will be neither distinguishable nor indistinguishable from getting x-spin down (he will not have a determinate belief concerning whether his second and third results agree); rather, the disjunctive experience will be indistinguishable from either x-spin up *or* x-spin down (the observer *will* determinately believe that the result of the third measurement *is indistinguishable from exactly one of the first two measurement results*, but again he will not have any determinate belief regarding *which* of the first two results it is indistinguishable from).

Further, if an observer can correctly identify those experiences that are ordinary and have specific determinate content when he has them, then it follows from the linear dynamics (and the assumption that the observer actually believes whatever he has a sure-fire disposition to report that he believes) that he would believe that his disjunctive experiences are perfectly ordinary and specific. If the observer has the sure-fire disposition to report 'Sharp!' when he determinately records x-spin up, and if he has the sure-fire disposition to report 'Sharp!' when he determinately records x-spin down, then he will have the sure-fire disposition to report 'Sharp!' when he is in a superposition of recording x-spin up and x-spin down. And this is why the observer would believe that all of his experiences were perfectly determinate and had ordinary specific content when he would in fact have no determinate belief concerning *which* sharp result he got.

Disjunctive experiences and beliefs have no ordinary specific content. It is as if they were the shells of ordinary experiences and beliefs with nothing inside. But it is precisely this lack of ordinary specific content that an observer would be unable to detect through introspection. If the bare theory were true, then first-person authority concerning whether particular experiences and beliefs had ordinary specific content would be routinely violated in a striking way—an observer would typically believe that he had an ordinary determinate experience when there would in fact be no such experience that he believed that he had.

One might be tempted to conclude that the bare theory simply fails to account for our experience since it predicts that our experience typically has no ordinary specific content when we know *by direct introspection* that it does. But I do not think that this is the right conclusion. If one believes that one's experience typically has ordinary specific content and that this is something about which one cannot possibly be mistaken, then one will obviously not like the account of experience provided by the bare theory. But it is also true that if one believes that a spoon bends as it is put into a glass of water and that this is something about which one cannot possibly be mistaken, then one will not like the account of experience provided by classical optics. The point is just that we typically give our theories the chance to explain why things are not as they seem. Just as classical optics explains why spoons seem to bend in water and why looking at a half-submerged spoon is an unreliable way to determine whether or not it is straight, the bare theory explains why disjunctive experiences seem to have ordinary determinate content when they in fact do not and why direct introspection is an unreliable way to determine whether an experience is ordinary or disjunctive. As with any theory, whether the bare theory provides a satisfactory account of our experience depends on precisely what one wants explained. In certain special situations (like those we have been discussing) the bare theory explains why one's experience seems to be ordinary and determinate when it is not, and, while it is uncomfortable to sacrifice first-person authority to this extent, it seems to me that this really might be taken as a satisfactory, though perverse, account of one's experience in these situations.

While the bare theory's account of the determinateness of experience is akin to the standard account of why spoons seem to bend in water, the illusions predicted by the bare theory are clearly more troubling than those predicted by classical optics. Indeed, the illusions predicted by the bare theory are so radical and would have to be so pervasive that they would ultimately undermine whatever empirical reasons one might think one had for accepting the bare theory in the first place. And it seems to me that this is the real problem.

The main point of this section is that while Everett thought that pure wave mechanics made the same empirical predictions as the standard collapse theory, if by pure wave mechanics he meant the bare theory, then he was wrong. While the standard theory predicts that I will typically get a single, determinate, and perfectly ordinary measurement result, the bare theory predicts that I will get a disjunctive result, which I will judge to be indistinguishable from *some* determinate result but which will fail to be determinately distinguishable or indistinguishable from any *particular* ordinary result. More specifically, while the standard theory might predict that an observer would get the ordinary result *up* or that he would get the ordinary result *down*, the bare theory would predict that the observer would believe that his result was perfectly ordinary and that it was *up* or *down*, but the observer would not determinately believe that he got *up* and not determinately believe that he got *down*. While an observer in the bare theory would believe that he got a determinate result, there would, contrary to the predictions of the standard theory, be no determinate result that he believed that he got. This means that if Everett had the bare theory in mind, then he did not deduce the standard theory's empirical predictions as subjective appearances.

4.5 *Problems with the bare theory*

One might take the fact that the bare theory makes empirical predictions of a fundamentally different sort than the standard theory, that it predicts disjunctive rather than ordinary experience, as a sufficient reason for judging that the bare theory is false. But that a theory predicts a new sort of illusion cannot by itself be a sufficient reason to abandon the theory, and it certainly does not provide any evidence that the theory is false—the physical world really could be deeply illusory. On the other hand, one should not conclude that all is well with the bare theory.

4.5.1 *Preferred basis*

Jeffrey Bub, Rob Clifton, and Brad Monton (BCM) have argued that the bare theory requires one to choose a preferred basis and that this undermines one of the most compelling arguments for the bare theory: its formal simplicity.

As we have seen, the bare theory accounts for reports of determinate experience by predicting that an observer would typically mistake disjunctive results for ordinary results. One might naturally wonder whether there is some way to wire an observer so that he would always provide reliable answers concerning the state of his memories of past observations. It turns out that the answer is *no*. The reason is that the bare theory places a very strong constraint on the reliability of an observer: if an observer must answer a question the same way in two orthogonal states but differently in an superposition of the orthogonal states in order to answer the question reliably, then no observer can answer the question reliably in general since, by the linearity of the dynamics, he will always answer the question the same way in the superposition as he does in the orthogonal states.

This is where the preferred basis comes in. BCM argue that since there is no universally reliable observer in the bare theory, one must choose the circumstances under which a good observer will be able to answer reliably a particular question about his own memory state and that this requires one to choose a preferred basis.

Suppose we want to be sure that some observer N has the disposition to answer the question 'Did you get a determinate result to your x-spin measurement?' with 'No' if he ends up in the state

$$|\psi_1\rangle = \frac{1}{\sqrt{2}}(|\uparrow\rangle_N|\uparrow\rangle_S + |\downarrow\rangle_N|\downarrow\rangle_S) \tag{4.22}$$

and 'No' if he ends up in the state

$$|\psi_2\rangle = \frac{1}{\sqrt{2}}(|\uparrow\rangle_N|\uparrow\rangle_S - |\downarrow\rangle_N|\downarrow\rangle_S). \tag{4.23}$$

If the system recording the memory of the initial measurement result is larger than a few particles, then it would be very difficult to wire N to perform this sort of introspection (since he would have to perform something akin to an interference experiment on the record in order to answer this question reliably in these states); but suppose that we succeed in wiring him accordingly. While N would reliably report whether he recorded a determinate x-spin result in states $|\psi_1\rangle$ or $|\psi_2\rangle$, he would mistakenly report that he failed to record a determinate result when he was in fact in an eigenstate of recording a determinate result, say x-spin up. If asked whether he recorded a determinate result, N would answer 'No' in state ψ_1 and 'No' in state $|\psi_2\rangle$, so by the linearity of the dynamics, he would also (though this time mistakenly) answer 'No' in state $|\uparrow\rangle_N|\uparrow\rangle_S = 1/\sqrt{2}(|\psi_1\rangle + |\psi_2\rangle)$.

Consider what happens when we try to tell the bare theory's determinate-result story with N. After he performs an ideal x-spin measurement of a system in an eigenstate of z-spin, rather than being under an illusion that he got a determinate result as in the standard sort of bare theory story, N would (correctly) believe that he did not get a determinate result. But again, N would also (mistakenly) report that he did not get a determinate report in those situations where he did. That is, if an observer were wired like N, then the observer would *never* believe that he got a determinate measurement result; and, of course, the standard bare theory stories are not going to work for an observer like this.

It is the fact that a proponent of the bare theory must thus choose one from an infinite number of possible ways that an observer might be wired that is supposed to worry us here (since choosing how an observer is wired amounts to choosing which of the observer's reports will be reliable, which amounts to choosing a preferred basis for the observer). If the only reason for choosing the standard bare theory model of an observer was to get the standard bare theory stories to work, then I think that there would be some reason to worry, but there are in fact several independent reasons one might want to model observers this way. First, we believe that we typically get determinate results, which in the context of the bare theory means that we are not wired like N. Secondly, we believe that if we are wired to answer any question reliably, then we are wired to answer the question 'Did you get a determinate result to your measurement?'

reliably when we either determinately did or determinately did not record a result. Thirdly, we have good physiological evidence that our brains are not wired to perform the sort of self-measurements that would reliably determine whether they are in a specific superposition of having recorded mutually incompatible results—our brains are either wired like N's to make complicated interference measurements on the brain subsystems that are used to record our memories or they are not, and what we know about brain physiology suggests that they are not. And finally, the standard bare theory model of observers is precisely Everett's model, so even if the model turned out to be wrong, it would at least be a historically faithful reading of Everett.

Another reason for not worrying too much about the preferred-basis problem in the bare theory is that the preferred basis plays a very modest role. In a many-worlds theory (Chapter 6), for example, the preferred basis determines what worlds there are and what physical properties are determinate in each world. In the bare theory the choice of the preferred basis has nothing to do with the fundamental nature of the physical universe or its dynamics but rather simply represents a plausible choice about how to model observers. I do not think that the preferred-basis problem is a very serious one for the bare theory, especially when compared to the other problems it faces.

4.5.2 *Empirical incoherence*

In a world described by the bare theory one would be fundamentally mistaken concerning even the basic nature of one's own experience. This radical failure of first-person authority is one of the things that makes the bare theory philosophically interesting, but it also leads to methodological problems. There is nothing inherently wrong with a theory predicting a failure of first-person authority, but because of the way that first-person authority fails in the bare theory, the theory ends up being empirically self-defeating—the truth of the bare theory would undermine whatever empirical justification one may have thought one had for accepting it in the first place because one would never be able to trust one's beliefs concerning what one had observed (or even *whether* one had observed anything determinate).

We can only accept a physical theory on empirical grounds if we can explain how, if it were in fact true, one might have empirical justification for accepting it. We will say that a theory that allows us to tell such a story is *empirically coherent*. If a theory fails to be empirically coherent, it might be true, but if true, then one could never have empirical reason for accepting it as true. In this sense, a theory that fails to be empirically coherent could not be an object of scientific knowledge. Again, such a theory might be true—it is just that if it were true, then one would never know that it was. The bare theory fails to be empirically coherent since it predicts that one would almost never be in a position to determine reliably the nature of one's own experience.

The bare theory requires one to be extraordinarily sceptical concerning the limits of knowledge. In his First Meditation Descartes considered the possibility that our experience might be the result of an evil demon bent on deception. He asks his reader to imagine a demon who could impress ideas on an observer's mind. If one's experience were the result of the actions of such a demon, then one would be justified in concluding very little about the actual nature of the physical world. But at least one would know what one had experienced, or at least what one was currently experiencing. The bare theory, however, does not even allow for this. Imagine a super-demon who can and typically does lead one to believe that one has determinate experiences with specific content when one's experience typically fails to be determinate and fails to have any ordinary specific content. With such a demon around, empirical inquiry would be impossible. And this is what it would be like for the bare theory to be true.[10]

4.5.3   *No account of statistical results*

There is not much to say about this one. The bare theory's general limiting property tells us what an observer would report in the limit as an infinite number of similar measurements were made on similarly prepared systems if one assumes that there is a well-defined limiting state; but we never in fact perform an infinite number of similar measurements on similarly prepared systems, so it is difficult to see how these limiting results could be relevant to our actual statistical reports and beliefs.

One might hope that knowing what an observer would report about his statistics in the limit would provide statistical information about *finite* sequences of measurement results. Ordinarily it might, but in the bare theory a finite system only determinately has a property when it is in an eigenstate of having a property. After a finite number of observations an observer will typically have no determinate sequence of measurement records nor anything determinate to say about the statistical properties of his results. He will, of course, believe that his results exhibit *some* determinate statistical properties (by the linearity of the dynamics and

---

[10] See Barrett (1996) for a more detailed discussion of empirical coherence.

because he would believe this of every determinate sequence of results), but he will typically not end up believing that they exhibit *any particular* statistical properties after a finite sequence of measurements. The bare theory thus has no explanation why the standard statistical predictions of quantum mechanics are in any way relevant to the experience of real observers.

Of course, the composite system will always be getting closer to a state where the observer would report that his first n results exhibit the usual quantum statistics. One might suppose that close is good enough and try weakening the standard eigenvalue–eigenstate link so that states close (in the appropriate Hilbert-space metric) to an eigenstate determinately have the corresponding physical property. This, however, has curious consequences.

Consider the experiment where M measures the same observable on a sequence of identically prepared systems. On the weakened eigenvalue–eigenstate link, the composite system would eventually be close enough to the appropriate eigenstate that one would be able to say that the observer has the sure-fire disposition to report and hence believes that his results were randomly distributed with the usual quantum relative frequencies. But by the same argument, since the norm-squared of the coefficient on each term in the determinate-record expansion goes to zero as the number of measurements gets large, the composite system would eventually be close enough to the appropriate eigenstates that one would be able to say that the observer would believe that he failed to get each particular length-n sequence of measurement results. Yet he would also be in a state where he would believe that he got *some* determinate sequence of results. That is, after a finite number of observations, the observer would believe that he got a determinate length-n sequence of results but that he did not get any particular length-n sequence of results, which, on most accounts, is a logical contradiction. So the weakened eigenvalue–eigenstate link together with the way that we have supposed that observers are wired would require an observer eventually to report and believe a logical contradiction on the basis of his most careful empirical observations![11]

---

[11] Some of these considerations are from conversations with Brad Monton. Note there is also a corresponding limiting argument in the bare theory with the standard eigenvalue–eigenstate link: in the limit as an observer makes an infinite number of measurements the composite system will approach a state where he reports that he failed to get each possible infinite sequence of results but also believes that he got some infinite sequence of results!

4.5.4 *No general account of determinate results*

Perhaps the best reason to reject the bare theory is that the standard bare theory stories only work if one supposes that an observer typically begins determinately sentient and ready to make measurements on systems that are somehow initially prepared in separable states, and the observer must stay determinately sentient throughout the experiment. If the bare theory were true, however, such just-right conditions would virtually never be met.

Suppose one sets out to measure the x-spin of a system in an eigenstate of z-spin and that it just happens that a large truck is driving by one's laboratory at the same moment. If the truck begins determinately outside the laboratory, then the usual linear dynamics tells us that it will not stay there for long; rather, it will very quickly evolve to a superposition of travelling along many mutually incompatible trajectories, and some of these trajectories would no doubt send it ploughing through the laboratory. Even if the coefficient on the term that describes the truck as crushing the observer is very small, if it is nonzero, then the observer will end up in a superposition of recording x-spin up, recording x-spin down, and recording nothing because he was run over by a truck just as he was about to glance at his measurement device. According to the standard eigenvalue–eigenstate link, such an observer would not end up determinately reporting or believing anything at all because he would no longer determinately be the sort of physical system that *could* make reports or have beliefs.

But the situation is actually worse than it may sound here since there would most likely fail to be any determinate sentient observer (or determinate truck) *before* the experiment either. As Albert put it:

If the bare theory is true, then it seems extraordinarily unlikely that the present quantum state of the world can possibly be one of those in which there's even a matter of fact about whether or not any sentient experimenters exist at all. And of course in the event that there *isn't* any matter of fact about whether or not any sentient experimenters exist, then it becomes unintelligible to inquire ... about what sorts of things experimenters will *report*. (1992: 124–5)

The point here is that one would expect a typical global quantum-mechanical state to be nowhere near being an eigenstate of there being trucks with determinate positions or observers with determinate beliefs (in the Hilbert-space metric and sense of typical), and without special initial conditions or special energy properties (neither of which are assumed by the bare theory), the bare theory would do nothing that would lead to an atypical global state.[12] The upshot is that if the bare theory were true, then there would typically be no determinate experience to explain nor anyone determinate to do the explaining. Such are the bare theory's problems.[13]

---

[12] Choose a fine-grained partition of the Hilbert space used to represent the global state so that each cell in the partition is assigned equal probability measure (in the Hilbert-space norm-squared measure). Choose a cell in the partition at random, then choose any state in that cell. Such a state counts as a typical global state.

[13] But that the global state is most likely nowhere near being an eigenstate of any physical property that interests us is something that any successful no-collapse theory must somehow take into account!

5

SELECTING A BRANCH

O N E problem we saw with the bare theory was that, in so far as it predicts determinate experiences, it tells us that they are typically disjunctive, and that they thus fail to have any ordinary determinate content. An obvious strategy for fixing the bare theory would be to drop the standard eigenvalue–eigenstate link and supplement the usual quantum-mechanical state with something that would provide observers with ordinary measurement results—even in those situations where they do not begin in an eigenstate of being ready to perform a particular measurement. One would then want to provide a dynamics for this extra aspect of the physical state.[1]

Here I shall consider how one might add a physical quantity to the usual quantum-mechanical state description in order to account for our determinate measurement records. The value of this quantity in effect selects exactly one Everett branch as actual—the branch that describes an observer as recording the measurement results that he in fact records. There are essentially two ways of doing this: (1) one might stipulate a single preferred physical quantity as always determinate, something like particle configuration in Bohm's theory, or (2) one might stipulate a rule, like the rules provided by various modal theories that are designed to select a determinate property for each physical system (hence for each observer) at each time given the current quantum-mechanical state. In either case, the complete description of the physical state at a time is given by the usual quantum-mechanical state *together with the value of the determinate physical quantities.*

---

[1] One might, as in a many-worlds theory, suppose that there is one world for each term in the global state written in the determinate-record basis, then provide a dynamics for how the local states of such worlds evolve (by stipulating a rule for determining which worlds at different times are in fact the same world). Or perhaps the most direct way to guarantee that experiences have ordinary determinate content would be, as in a many-minds theory, to add the mental states of observers to the usual quantum-mechanical state, then provide a dynamics for the mental states given the time-evolution of the wave function. I shall consider each of these proposals later (in Chs. 6 and 7, respectively).

On this sort of theory, what the determinate quantities are and how their values change over time is supposed to explain why we record what we in fact record. In order for a modal or hidden-variable theory to account for our experiences and beliefs, then, one must be convinced that it is precisely those physical quantities that the particular theory makes determinate that determines our experiences and beliefs—that is, one must believe that the determinate quantities make our most immediately accessible physical records determinate. On Bohm's theory, for example, the positions of particles are always determinate: if this makes our most immediately accessible physical records determinate, then Bohm's theory provides an account of our determinate experience; if not, then it is no better off than the bare theory.

Any theory that selects a single Everett branch as privileged is incompatible with Everett's claim that the usual quantum-mechanical state provides a complete physical description. The usual state determines all possible branches on all possible resolutions of the state; it does not select any branch as *actual*. But perhaps one can only get a satisfactory no-collapse theory by violating this principle of state completeness. It ultimately depends on precisely what one wants from a satisfactory theory.

5.1   *Bohm's theory without the trajectories*

The formulation of quantum mechanics that David Bohm described in 1952 has long stood as a counter-example to many of the deep philosophical conclusions that people have tried to draw from quantum mechanics. According to Bohm's theory the world consists of a collection of particles that always have determinate positions and always follow continuous trajectories that are determined by the usual linear evolution of the wave function. While Bohm's theory is deterministic, it none the less makes the same statistical predictions as the standard collapse theory for particle positions (probabilities in Bohm's theory are naturally understood epistemically) whenever the standard theory makes coherent predictions. Unlike the standard theory, however, Bohm's theory treats measurements the same way it treats every other physical interaction—a system follows precisely the same dynamics during a measurement as it does the rest of the time. Measurements have determinate outcomes on Bohm's theory because every particle always has a determinate position and it is assumed, as a basic interpretational principle of the theory, that the result of every measurement is ultimately recorded in terms of the position of something.[2]

While Bohm's theory is often called a hidden-variable theory, where the positions of particles are the so-called hidden variables, this is rather misleading. As Bohm and Hiley put it:

> our variables are not actually hidden. For example, we introduce the concept that the electron *is* a particle with a well-defined position and momentum that is, however, profoundly affected by a wave that always accompanies it...Far from being hidden, this particle is generally what is most directly manifested in an observation. (Bohm and Hiley 1993: 2)

Again, it is the determinate particle positions that are supposed to account for our determinate experiences and beliefs on Bohm's theory.

But before considering how Bohm's theory works, I shall consider a more primitive theory. Bell described his interpretation of Everett, his Everett (?) theory, as Bohm's theory without the continuous particle trajectories:

> instantaneous classical configurations are supposed to exist, and to be distributed in the comparison class of possible worlds with probability $|\psi|^2$. But no pairing of configurations at different times, as would be effected by the existence of trajectories, is supposed. And it is pointed out that no such continuity between present and past configurations is required by experience. (Bell 1987: 133)

The wave function ψ evolves in a perfectly linear way, but the classical configuration (the positions of each particle) jumps around in a random way that depends only on $|\psi|^2$. This means that one's measurement records would typically change in a pathologically discontinuous way over time and hence be wildly unreliable as records of what actually happened.

Consider how Bell's Everett (?) theory would describe an x-spin measurement. Suppose that an observer M measures the x-spin of a spin-$\frac{1}{2}$ system S that is in a superposition of different x-spins. The composite system will end up in the familiar superposition of M recording x-spin up

---

[2] It is sometimes said that in Bohm's theory one assumes that every measurement is ultimately a measurement of position, but this, I believe, just means that one assumes that every measurement is ultimately recorded in the position of something. One can get even clearer by noting that the content of a measurement record in Bohm's theory is determined by the position of something *relative* to the wave function—that is, a different wave function and the same position might produce a different record. I believe that this point is at least implicit in Albert's (1992) discussion of how measurement works in Bohm's theory.

and S being x-spin up and M recording x-spin down and S being x-spin down:

$$\alpha|\uparrow\rangle_M|\uparrow\rangle_S + \beta|\downarrow\rangle_M|\downarrow\rangle_S;\qquad(5.1)$$

but, on the Everett (?) theory, *the particle configuration is always determinate*. This means that if M records his measurement result in terms of the positions of particles, he will end up with a determinate record corresponding to one or the other of the two terms in the final state—that is, the particle configuration will effectively select one of the two terms in (5.1) as actual and M is thus guaranteed to end up with the determinate record *x-spin up* or the determinate record *x-spin down*. Which configuration the observer ends up with, hence which determinate physical record he gets, is randomly determined, where the probability of M getting a particular record is equal to the norm-squared of the amplitude of the wave function associated with the record. If the amplitudes associated with the records *x-spin up* and *x-spin down* are α and β respectively (as above), then the probability of the observer ending up with a configuration recording x-spin up is $|\alpha|^2$ and the probability of the observer ending up with a configuration recording x-spin down is $|\beta|^2$.

Suppose that the observer gets the result x-spin up for the outcome of his first measurement. Now what happens if the observer carefully repeats his measurement? The usual linear dynamics tells us that the state after the second measurement will be

$$\alpha|\uparrow,\uparrow\rangle_M|\uparrow\rangle_S + \beta|\downarrow,\downarrow\rangle_M|\downarrow\rangle_S.\qquad(5.2)$$

So what is the observer's actual memory configuration? Again, the probability of the actual configuration ending up associated with a particular term in the quantum-mechanical state is completely determined by the current quantum-mechanical state and is independent of past configurations. It is given by the square of the coefficients on the terms when the wave function is written in the configuration basis. Here, then, there is a probability of $|\beta|^2$ that M will end up with a configuration recording x-spin down for the second result even though he actually recorded x-spin up for the first result. In other words, there is a probability of $|\beta|^2$ that the observer's second measurement result will *disagree* with his first. But if M does in fact get x-spin down for his second measurement, the classical configuration will now be one associated with the second term of the above state, which means that M's 'record' of his first measurement (in terms of the configuration) will now be x-spin down, and it will thus *appear*, based on an examination of his records, that the two measurements did in fact yield the same result. Because the sum of the norm-squared amplitudes associated with those classical configurations where the observer's records exhibit the usual quantum statistics would typically be close to one, the actual configuration would almost always be such that the observer's records would exhibit the usual quantum statistics and correlations. So it would typically *appear* that the standard collapse theory made the right empirical predictions even though the observer's memories would in fact be wildly unreliable (Fig. 5.1).

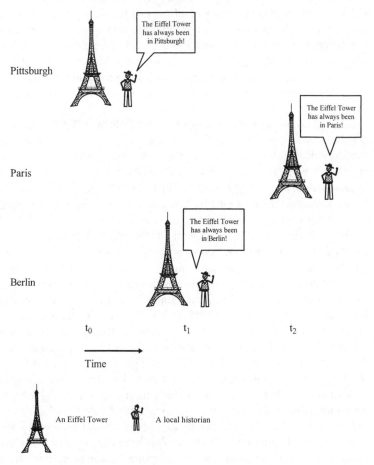

FIG. 5.1 A relatively tame Eiffel-Tower trajectory on Bell's Everett (?) theory.

As Bell put it:

in our interpretation of the Everett theory there is no association of the particular present with any particular past. And the essential claim is that this does not matter at all. For we have no access to the past. We have only our 'memories' and 'records'. But these memories and records are in fact *present* phenomena. . . . The theory should account for the present correlations between these present phenomena. And in this respect we have seen it to agree with ordinary quantum mechanics, in so far as the latter is unambiguous. (Bell 1987: 135–6)

Bell thought of the particle configuration at a time as selecting the single Everett branch that describes the actual physical configuration of our world, and he thought that one only needs one world to account for our experience—*our world* (Bell 1987: 97, 133). One could, however, convert Bell's Everett (?) theory into a many-worlds theory by imagining an infinite collection of worlds, each with a determinate particle configuration and each evolving randomly according to the dynamics Bell describes.

So what are we to make of Bell's Everett (?) theory? Bell for one did not like it much. He said that 'if such a theory were taken seriously it would hardly be possible to take anything else seriously' (1987: 136; see also 1987: 98). He seems to have been worried most about the social implications of the theory; but one might also worry that if the Everett (?) theory were true, then, like the bare theory, one could have no empirical reason for accepting it as true.

The problem is that Bell's Everett (?) theory, though arguably better off than the bare theory, is not empirically coherent. While one can test the instantaneous empirical predictions of the Everett (?) theory (the way that records are correlated at a particular instant), because one's memories are wildly unreliable one can have no empirical justification for accepting any specified dynamical law (for how the particle configuration evolves). In particular, if the Everett (?) theory were true, then one could have no empirical justification for accepting its pathologically discontinuous dynamics as the correct description of the time-evolution of the particle configuration. Another way to put this is that if one is willing to concede that one's memories concerning past configurations were unreliable in the sense required to take the Everett (?) theory seriously, then one would have exactly as much empirical justification for accepting that the physical state has always been exactly what it is *right now* as one would have for accepting the dynamics given by the Everett (?) theory.

One could make the Everett (?) theory empirically coherent by changing its dynamics so that measurement records are typically reliable records of past events. Bohm's theory provides just such a dynamics.

5.2 *Bohm's theory*

On Bohm's theory, the quantum-mechanical state ψ evolves in the usual linear, deterministic way, but particles always have determinate positions and move along continuous trajectories described by an auxiliary dynamics. In its simplest form Bohm's theory is characterized by a few basic principles:[3]

1. *State description.* The complete physical state at a time is given by the wave function ψ and the determinate particle configuration Q.

2. *Wave dynamics.* The time-evolution of the wave function is given usual linear dynamics. In the simplest situations, this is just Schrödinger's equation:

$$i\hbar\frac{\partial\psi}{\partial t} = \hat{H}\psi. \tag{5.3}$$

More generally, one uses the form of the linear dynamics appropriate to one's application (as in the x-spin examples discussed below).

3. *Particle dynamics.* The particles move according to

$$\frac{dQ_k}{dt} = \frac{1}{m_k}\left.\frac{\text{Im}(\psi^*\nabla_k\psi)}{\psi^*\psi}\right|_Q, \tag{5.4}$$

where m_k is the mass of particle k and Q is the current configuration.

4. *Distribution postulate.* There is a time t_0 where the epistemic probability density for the configuration Q is given by $\rho(Q, t_0) = |\psi(Q, t_0)|^2$.

If there are N particles, then ψ is a function in $3N$-dimensional configuration space (three dimensions for the position of each particle), and the current particle configuration Q is represented by a single point in configuration space (in configuration space a single point gives the determinate position of every particle). Each particle moves in a way that depends on its position, the evolution of the wave function, and the positions of every other particle. The easiest way to picture the particle dynamics is to picture the point representing the N-particle configuration as being carried

---

[3] This description of Bohm's theory follows Bell (1987: 127) rather than Bohm's quantum-potential description.

along by the probability currents generated by the linear evolution of the wave function ψ *in configuration space.*

Bell put this point in a rather dramatic way: '*No one can understand this theory until he is willing to think of ψ as a real objective field rather than just a "probability amplitude." Even though it propagates not in 3-space but in 3N-space*' (1987: 128). While Bell's position here might be a bit extreme, it really does seem that the natural place to tell causal stories in Bohm's theory is in configuration space. In any case, since the dynamics is deterministic, possible trajectories never cross at a time in configuration space.

Since every particle always has a determinate position even when the universal wave function fails to be in an eigenstate of configuration, Bohm's theory violates the standard eigenvalue–eigenstate link. Indeed, Bohm's theory does not use either direction of the standard interpretation of states: not being in an eigenstate of position does not mean that a system fails to have a determinate position; and being in an eigenstate of some observable, x-spin say, is typically not understood to mean that the system has a determinate x-spin. Rather, since position is the only noncontextual, observable property, it is usually taken to be the only intrinsic, observable property that a particle can have.[4] Consequently, one must assume that every measurement is ultimately a measurement of position; or better, one must assume that the fact that particles always have determinate positions ultimately explains our determinate experiences and beliefs.[5]

Since the total particle configuration can be thought of as being carried around by the probability current in configuration space, the probability of the particle configuration being found in a particular region of configuration space changes as the integral of $|\psi|^2$ over that region changes. More specifically, the continuity equation

$$\frac{\partial \rho}{\partial t} + \mathrm{div}(\rho v^{\psi}) = 0 \tag{5.5}$$

is satisfied by the probability density $\rho = |\psi|^2$. And this means that if the epistemic probability density for the particle configuration is ever $|\psi|^2$, then it will always be $|\psi|^2$, unless one makes an observation.[6] That is,

---

[4] A contextual quantity is one whose value typically depends on how one goes about measuring it.

[5] One can empirically determine the *effective wave function* of a system, which is a function of the actual particle configuration and the global quantum state.

[6] See Dürr *et al.* (1993a) for a discussion of the equivariance of the statistical distribution ρ.

if one starts with an epistemic probability density of $\rho(t_0) = |\psi(t_0)|^2$, then, given the dynamics, one should update this probability density at time t so that $\rho(t) = |\psi(t)|^2$. And if one makes an observation, then the epistemic probability density will be given by the *effective* wave function.

Suppose one is prepared to measure the x-spin of a particle P in an eigenstate of z-spin, say $|\uparrow_z\rangle_P$. What it means for P to be in determinate z-spin state in Bohm's theory is that one has a roughly localized and symmetric wave packet in a quantum-mechanical state like

$$\frac{1}{\sqrt{2}}(|\uparrow_x\rangle_P + |\downarrow_x\rangle_P)|S\rangle_P, \tag{5.6}$$

where the first half of the expression is the spin component and the second is the position component (the wave function is initially concentrated in the start region S) and where the particle P is in fact somewhere in the wave packet with a probability density given by $|\psi|^2$. If such a wave packet were to pass through a Stern–Gerlach device oriented along the z-axis, then the whole wave packet would be deflected. But if such a wave packet were to pass through a Stern–Gerlach device oriented along the x-axis (as shown in Fig. 5.2), then the two spin components of the wave function will be separated: the $|\uparrow_x\rangle_P$ component will be deflected

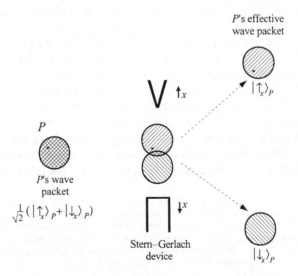

FIG. 5.2 How the probability current moves a single particle. Note that P does not move in the x-direction until it is influenced by only one spin component.

up and the $|\downarrow_x\rangle_P$ component will be deflected down (Fig. 5.2). By the distribution postulate, and by symmetry, there is an even chance that P begins in the top or bottom half of the initial wave packet. And if P begins in the top half of the wave packet, then it will be carried up by the probability current; and if it begins in the bottom half of the wave packet, then it will be carried down. Further, if it is in the top half of the top half of the initial wave packet, then it will end up in the top half of the up wave packet; if it is in the bottom half of the top half of the initial wave packet, then it will end up in the bottom half of the up wave packet; if it is in the top half of the bottom half of the initial wave packet, then it will end up in the top half of the down wave packet; and finally, if it is in the bottom half of the bottom half of the wave packet, then it will end up in the bottom half of the down wave packet. Whichever wave packet P ends up in will determine how it will behave on a second x-spin measurement (if, for example, it ends up associated with the spin up component, then it will surely be deflected up on a subsequent x-spin measurement since all of the probability current will flow up). On the other hand, the shifting in the relative position of the particle in the wave packet is what accounts for the fact that, if the distribution postulate is satisfied, then one will not know how a second x-spin measurement will come out after a subsequent z-spin measurement.[7]

Trajectories cannot cross in configuration space in Bohm's theory, and here configuration space is ordinary 3-space. Also, note that the direction that P is deflected in this experiment is as much determined by the state of the measuring device as by the state of the system being measured. More specifically, one would get exactly the opposite x-spin result for a particular trial if one modified the Stern–Gerlach device so that it deflects the $|\uparrow_x\rangle_P$ component *down* and the $|\downarrow_x\rangle_P$ component *up*.[8] If the particle was deflected up, for example, then it was in the top half of the initial wave packet and thus it would also be deflected up by the modified Stern–Gerlach device but it would now behave like an x-spin down particle instead of an x-spin up particle. It is because the spin behaviour that a

[7] Since the result of such an x-spin measurement would then depend not on whether P was initially in the top or bottom half of the wave packet before the first experiment, which is something one could know before the second x-spin measurement, but, rather, on whether P was initially in the top half of the top half, bottom half of the top half, top half of the bottom half, or bottom half of the bottom half before the first x-spin measurement, which according to the distribution postulate is something one would not know before the second x-spin measurement.

[8] Unlike Albert's (1992) colour and hardness boxes, this modification would involve more than just rotating our measuring device by 180 degrees.

particle exhibits is determined by the details of how one performs one's measurement (the same particle is determinately up for one device but determinately down for another), spin properties are *contextual*, and thus are not usually considered to be genuine physical properties on Bohm's theory. Again, since position is the only noncontextual property, it is usually taken to be the only fundamental physical property.

Let's return to our first Stern–Gerlach device, and suppose we record the direction that P is deflected in the position of another system, call it M. Suppose that the interaction is such that if P were in a pure x-spin up wave packet, then M's wave packet would not move; and if P were in a pure x-spin down wave packet, then M's wave packet would be deflected down (Fig. 5.3(a)). By the linearity of the wave dynamics, then, the two-particle wave function after the interaction will be

$$\frac{1}{\sqrt{2}}(|\uparrow_x\rangle_P |\text{up}\rangle_P |\text{records up}\rangle_M + |\downarrow_x\rangle_P |\text{down}\rangle_P |\text{records down}\rangle_M).$$

(5.7)

And M will record up if and only if P is deflected up. Here ψ evolves in $6N$-dimensional configuration space and one must keep track of the point representing the two-particle configuration (Fig. 5.3(b)). The position of M tells us which wave packet the two-particle configuration ends up associated with, but the record does not tell us precisely *where* the configuration is within that wave packet. While we know how P will behave on a subsequent x-spin measurement, we do not, for example, know how it will behave on a subsequent z-spin measurement or a second x-spin measurement if that measurement follows a z-spin measurement.

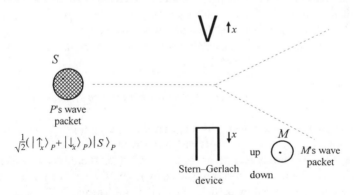

FIG. 5.3(a) A simple recording in Bohm's theory. M's wave packet moves to *down* if and only if P travels the $\downarrow_x$-path.

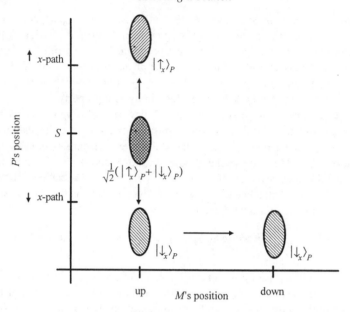

FIG. 5.3(*b*) The two-particle configuration in a two-particle wave packet. NOTE
that *M* moves to *down* if and only if the two-particle configuration starts in the
bottom half of this two-particle wave packet.

More specifically, when we condition on the information about *P*'s position gained from *M*, we can rule out one or the other of the two resulting
wave packets as determining *P*'s future behaviour, but we cannot tell
where *P* is in its new *effective wave packet* ψ_E. If ρ is the epistemic
probability density and ψ is the global wave function, then it is no longer
the case that $\rho = |\psi|^2$; rather, after conditioning on the new information
provided by the position of *M*, $\rho = |\psi_E|^2$.

5.3   *Surreal trajectories and the persistence of memory*

What makes Bohm's theory better than Bell's Everett (?) theory is that
the particle trajectories connect past events with current records in such a
way that our current records are typically reliable. There is an argument,
however, that the trajectories predicted by Bohm's theory are not the real
trajectories followed by particles, but rather are 'surreal'.[9] The worry is

[9] The original argument and this terminology comes from Englert *et al.* (1992).

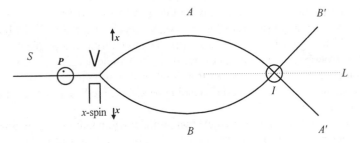

FIG. 5.4 Crossing-paths setup.

that one might observe a track in a bubble chamber when Bohm's theory predicts that the particle in fact travelled a very different path. If Bohm's theory really did predict the wrong trajectories for particles, then this would be a problem (the particle dynamics would be false and, among other things, we are relying on the particle dynamics to guarantee the empirical coherence of Bohm's theory). I believe, however, that so-called surreal trajectories pose no serious problems for Bohm's theory, and I will argue this in the context of the minimal version of Bohm's theory described earlier.

Consider an experiment where a spin-$\frac{1}{2}$ particle P starts at region S in a z-spin up eigenstate, has its wave packet split into an x-spin up component that travels from A to A' and an x-spin down component that travels from B to B' (Fig. 5.4).[10] The wave function evolves as follows:

Initial state:

$$|\uparrow_z\rangle_P|S\rangle_P = 1/\sqrt{2}(|\uparrow_x\rangle_P + |\downarrow_x\rangle_P)|S\rangle_P. \tag{5.8}$$

After the initial wave packet splits:

$$1/\sqrt{2}(|\uparrow_x\rangle_P|A\rangle_P + |\downarrow_x\rangle_P|B\rangle_P). \tag{5.9}$$

Final state:

$$1/\sqrt{2}(|\uparrow_x\rangle_P|A'\rangle_P + |\downarrow_x\rangle_P|B'\rangle_P). \tag{5.10}$$

By symmetry, the probability current across the line L is always zero. This means that if P starts in the top half of the initial wave packet, then it will move from S to A to I to B'; and if it starts in the bottom half of

---

[10] Since the standard line is that position is the only observable physical quantity in Bohm's theory, this does not mean that P has a determinate z-spin; rather, it is just a description of the spin index associated with P's effective wave function.

the initial wave packet, then it will move from S to B to I to A'. In other words, one might say that whichever path P takes, it will 'bounce' when it gets to region I. The problem is that quantum mechanics predicts, and our experience confirms, that if we detect a particle at A, then it will be found at A'; and if we detect it at B, then it will be found at B'. That is, we never observe the odd bouncing behaviour predicted by Bohmian mechanics.

Since Bohm's theory predicts that particles can bounce out of regions where there are no fields (at least no fields in the usual sense of a field), one might try to use this fact to cause problems for the theory. This is precisely what Englert *et al.* (1992) try to do. Their argument goes something like this:

Assumption 1 (explicit). Our experimental measurement records tell us that in a two-path experiment like that described above each particle either travels from A to A' or from B to B'; that is, they never bounce.

Assumption 2 (implicit). Our measurement records reliably tell us where a particle is at the moment the record is made.

Assumption 3 (implicit). One can record which path a particular particle takes without breaking the symmetry in the probability currents that prevents the particle from crossing the line L.

Conclusion. The trajectory predicted by Bohm's theory, where the particle bounces, cannot be the particle's actual trajectory; that is, Bohm trajectories are not real, they are 'surreal'. And note that if the trajectories predicted by Bohm's theory are not the actual trajectories, then Bohm's theory is false, which would constitute very good grounds for not liking it.

Note that Englert *et al.* do not claim that Bohm's theory makes the wrong *empirical* predictions, nor do they claim that it is somehow logically inconsistent; rather, they argue (on the implicit assumption that our particle detectors reliably tell us where particles are) that Bohm's theory makes the wrong predictions for the actual motions of particles. Dewdney *et al.* (1993) have tried to defend Bohm's theory in the context of an experiment like that described by Englert *et al.* above by denying that their detectors would be reliable. While this is certainly an option for a proponent of Bohm's theory, I shall argue instead that whenever one makes a determinate record in Bohm's theory using a device that induces a strong correlation between the measured property and the position that records the outcome, that record will be reliable *at the instant the determinate record is made*. Rather than show that Bohm's theory is false

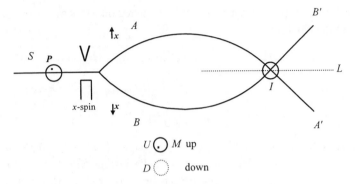

FIG. 5.5 Crossing paths with position record.

or that it requires one to deny the reliability of good measuring devices, one could think of experiments like those described by Englert *et al.* ultimately illustrating the inherently nonlocal nature of Bohm's theory.[11]

Consider what happens when one records the path travelled by the particle in terms of the position of a recording particle M (Fig. 5.5). That is, the interaction between P and M is such that if P's effective wave function was an up eigenstate of x-spin, then M's effective wave function would stay at U; and if P's effective wave function was a down eigenstate of x-spin, then M's effective wave function would move to record D. The effective wave function of the composite system evolves as follows:

Initial state:

$$|\uparrow_z\rangle_P |S\rangle_P |U\rangle_M = |S\rangle_P |U\rangle_M 1/\sqrt{2}(|\uparrow_x\rangle_P + |\downarrow_x\rangle_P). \qquad (5.11)$$

P's wave packet splits:

$$|U\rangle_M 1/\sqrt{2}(|\uparrow_x\rangle_P |A\rangle_P + |\downarrow_x\rangle_P |B\rangle_P). \qquad (5.12)$$

M's position is correlated with the position of P:

$$1/\sqrt{2}(|\uparrow_x\rangle_P |A\rangle_P |U\rangle_M + |\downarrow_x\rangle_P |B\rangle_P |D\rangle_M). \qquad (5.13)$$

---

[11] Actually, the experiments that they describe are not really in the domain of Bohm's theory since their 'one-bit detectors' would presumably require a field-theoretic description. But I think that their experiments can be translated into experiments within the domain of Bohm's theory (as described below).

The two wave packets appear to pass though region I (but they *miss* each other in configuration space!):

$$1/\sqrt{2}(|\uparrow_x\rangle_P|I\rangle_P|U\rangle_M + |\downarrow_x\rangle_P|I\rangle_P|D\rangle_M). \qquad (5.14)$$

Final state:

$$1/\sqrt{2}(|\uparrow_x\rangle_P|A'\rangle_P|U\rangle_M + |\downarrow_x\rangle_P|B'\rangle_P|D\rangle_M). \qquad (5.15)$$

So does the position of M reliably record where P was when the record was made? Yes. What happens here is that the position with M correlation destroys the symmetry that prevents P from crossing L: while the two wave packets both appear to pass though region I, they in fact miss each other in configuration space (Fig. 5.6).

In order to see how P and M move in this experiment, consider the evolution of the wave function and the two-particle configuration in configuration space. If the two-particle configuration starts in the top half of the initial wave packet, then P would move from S to A to I to A' and M would stay at U. If the configuration starts in the bottom half of the initial wave packet, then P would move from S to B, then M would move to D, then P would move from B to I to B'. That is, regardless of where P starts, it will pass though the region I (without bouncing). Moreover, M will record that P was on path A if and only if P ends up at A' and that P was on path B if and only if P ends up at B'. That is, when we make a determinate record of P's position before P gets to I, then P

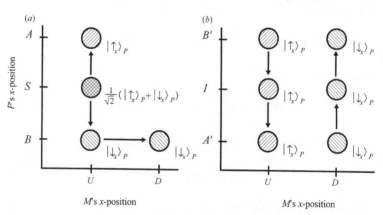

FIG. 5.6 The same in configuration space. (*a*) First half of the experiment. (*b*) Second half of the experiment.

will follow a perfectly natural trajectory, and our record will be reliable. Again, recording the position of P destroys the symmetry that prevents P from crossing L.

Records in terms of position are thus perfectly reliable in Bohm's theory if there is a strong correlation between the position of the system being observed and the position of the recording system. And since position is the only determinate physical quantity on Bohm's theory, one might simply conclude that all determinate records produced by strong correlations are reliable in Bohm's theory and dismiss the surreal trajectories objection altogether. This is not such a bad conclusion, but I think that the right thing to say about surreal trajectories is a bit more subtle.

Consider an experiment where one *tries* to record P's position in something other than the position of M. One might naturally, and quite correctly, object that there is no other determinate quantity in Bohm's theory that one *could* use to record P's position, but in the spirit of trying to revive Englert *et al.*'s challenge to Bohm's theory, read on. Suppose, for example, that one tries to record P's position in M's x-spin (and while one would not talk this way in Bohm's theory, it is still clear enough what this means): suppose that the interaction between P and M is such that if P's initial effective wave function were an x-spin up eigenstate, then nothing would happen to M's effective wave function; but if P's initial effective wave function were an x-spin down eigenstate, then the spin index of M's effective wave function would be flipped from x-spin up to x-spin down (Fig. 5.7). In the standard formulation of quantum mechanics one would naturally think of this interaction as recording P's position in M's x-spin (once a collapse had eliminated one term or the other of the superposition).

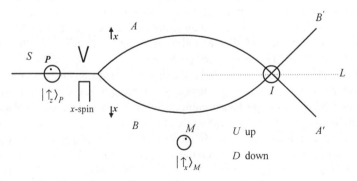

FIG. 5.7 We *try* to record P's position in M's x-spin.

The effective wave function of the composite system evolves as follows:

Initial state:

$$|\uparrow_z\rangle_P|S\rangle_P|\uparrow_x\rangle_M|R\rangle_M = |S\rangle_P|\uparrow_x\rangle_M|R\rangle_M 1/\sqrt{2}(|\uparrow_x\rangle_P + |\downarrow_x\rangle_P). \tag{5.16}$$

P's wave packet is split:

$$|\uparrow_x\rangle_M|R\rangle_M 1/\sqrt{2}(|\uparrow_x\rangle_P|A\rangle_P + |\downarrow_x\rangle_P|B\rangle_P). \tag{5.17}$$

The x-spin component of M's wave packet is correlated to the position of P's:

$$|R\rangle_M 1/\sqrt{2}(|\uparrow_x\rangle_P|A\rangle_P|\uparrow_x\rangle_M + |\downarrow_x\rangle_P|B\rangle_P|\downarrow_x\rangle_M). \tag{5.18}$$

The two wave packets pass through each other in configuration space:

$$|R\rangle_M 1/\sqrt{2}(|\uparrow_x\rangle_P|I\rangle_P|\uparrow_x\rangle_M + |\downarrow_x\rangle_P|I\rangle_P|\downarrow_x\rangle_M). \tag{5.19}$$

Then they separate:

$$|R\rangle_M 1/\sqrt{2}(|\uparrow_x\rangle_P|A'\rangle_P|\uparrow_x\rangle_M + |\downarrow_x\rangle_P|B'\rangle_P|\downarrow_x\rangle_M). \tag{5.20}$$

Then M's position is correlated to the x-spin component of its wave packet:

$$1/\sqrt{2}(|\uparrow_x\rangle_P|A'\rangle_P|\uparrow_x\rangle_M|U\rangle_M + |\downarrow_x\rangle_P|B'\rangle_P|\downarrow_x\rangle_M|D\rangle_M). \tag{5.21}$$

Note that here the symmetry in the probability current that prevents P from crossing L is preserved (Fig. 5.8). That is, P bounces just as it did in the first experiment. If the configuration begins in the top half of the initial wave packet, then P will move from S to A to I to B', then M will move to D. If the two-particle configuration begins in the bottom half of the initial wave packet, then P will move from S to B to I to A', then M will move to U. That is, the final position of M will be at U if P travelled along the lower path and it will be at D if P travelled along the upper path. In other words, M's final position does not reliably tell us where P *was* when the two wave packets became correlated. It does, however, reliably tell us where P *is* at the moment that the spin correlation was converted into a determinate measurement record (when the x-spin correlation was converted into a position correlation)—specifically, P is associated with the same two-particle spin packet as M wherever this requires P to be in ordinary 3-space.

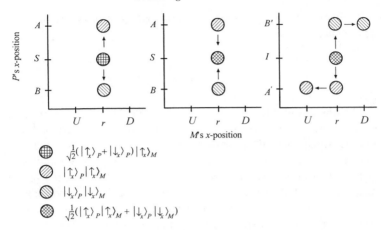

$\bigoplus$ $\quad \frac{1}{\sqrt{2}}(|\uparrow_x\rangle_P + |\downarrow_x\rangle_P)|\uparrow_x\rangle_M$

$\oslash$ $\quad |\uparrow_x\rangle_P |\uparrow_x\rangle_M$

$\obot$ $\quad |\downarrow_x\rangle_P |\downarrow_x\rangle_M$

$\otimes$ $\quad \frac{1}{\sqrt{2}}(|\uparrow_x\rangle_P |\uparrow_x\rangle_M + |\downarrow_x\rangle_P |\downarrow_x\rangle_M)$

FIG. 5.8 The same in configuration space.

Since the only determinate records in Bohm's theory are records in the position of something, M did not determinately record P's position until we converted the spin correlation into a position correlation. And when this position correlation is made, we reliably, and *nonlocally*, generate a record of P's position *at that moment*: If M moves to U, then this means that P is associated with the x-spin up component at A'; and if M moves to D, then this means that P is associated with the x-spin down component at B'.

The moral is that one cannot use a determinate record in Bohm's theory to figure out which path P took unless one knows *when* the record was made. But in this, Bohm's theory is far better off than the standard collapse theory since on the eigenvalue–eigenstate link one can say *nothing* about which trajectory a particle followed since it failed to have a determinate position until it was observed. If one does not worry about retrodiction in the context of the standard theory (and Englert *et al.* do not seem to worry about *this*), then there is certainly nothing to worry about in the context of Bohm's theory.

There is no reason that I can see, then, to suppose that the Bohm trajectories are not the actual particle trajectories. Nor is there any reason to conclude that our detectors are unreliable (one might say that the detector is not fooled by a late measurement but rather that it *reliably* detects the position of the test particle *nonlocally*). Whenever the position of one system is recorded in the position of another system via a strong correlation between the effective wave functions of the two particles, then that record will reliably tell one where the measured particle is *at the moment the determinate record is made.*

5.4   *The failure of covariance*

While surreal trajectories do not pose a serious problem for Bohm's theory, the way that the theory accounts for nonlocal statistical correlations does. In order to be compatible with relativity, a theory's dynamics must be Lorentz-covariant, and the dynamics that described particle motions in Bohm's theory is not. This is illustrated by the fact that space-like separate measurements will have different outcomes on the theory depending on the temporal order in which they are performed. The problem, of course, is that according to relativity theory there is no matter of fact about the temporal order of such events.[12]

Consider an EPR experiment in Bohm's theory. Two particles P_A and P_B begin in the EPR state

$$\frac{1}{\sqrt{2}}(|\uparrow_x\rangle_{P_A}|\downarrow_x\rangle_{P_B} - |\downarrow_x\rangle_{P_A}|\uparrow_x\rangle_{P_B}) \tag{5.22}$$

in some region S; then they move away from each other (say to regions A and B respectively) and x-spin measurements are made on each particle (Fig. 5.9). If the x-spin of particle P_A is measured first, then the state of the composite system will evolve as follows:

Initial state:

$$\frac{1}{\sqrt{2}}(|\uparrow_x\rangle_{P_A}|\downarrow_x\rangle_{P_B} - |\downarrow_x\rangle_{P_A}|\uparrow_x\rangle_{P_B})|S\rangle_{P_A}|S\rangle_{P_B}. \tag{5.23}$$

State after x-spin measurement on P_A at A:

$$\frac{1}{\sqrt{2}}(|\uparrow_x\rangle_{P_A}|A^+\rangle_{P_A}|\downarrow_x\rangle_{P_B} - |\downarrow_x\rangle_{P_A}|A^-\rangle_{P_A}|\uparrow_x\rangle_{P_B})|S\rangle_{P_B}. \tag{5.24}$$

State after x-spin measurement on P_B at B:

$$\frac{1}{\sqrt{2}}(|\uparrow_x\rangle_{P_A}|A^+\rangle_{P_A}|\downarrow_x\rangle_{P_B}|B^-\rangle_{P_B} - |\downarrow_x\rangle_{P_A}|A^-\rangle_{P_A}|\uparrow_x\rangle_{P_B}|B^+\rangle_{P_B}). \tag{5.25}$$

---

[12] The main argument of this section follows Albert (1992).

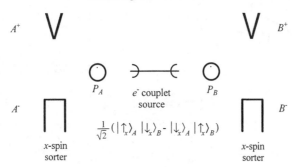

$$\frac{1}{\sqrt{2}}(|\uparrow_x\rangle_A |\downarrow_x\rangle_B - |\downarrow_x\rangle_A |\uparrow_x\rangle_B)$$

FIG. 5.9 An EPR experiment in Bohm's theory.

Suppose that each particle begins in the top half of the initial wave packet.[13] When P_A is measured at A, the wave packet will split into two components $|\uparrow_x\rangle_{P_A}|\downarrow_x\rangle_{P_B}$, which will go to region A^+, and $|\downarrow_x\rangle_{P_A}|\uparrow_x\rangle_{P_B}$, which will go to region A^-, and since the particle P_A begins in the top half of the initial wave packet, the probability current will carry it to region A^+ (Fig. 5.10(*a*)). The two-particle configuration will now be in a region of configuration space where the $|\downarrow_x\rangle_{P_A}|\uparrow_x\rangle_{P_B}$ component of the wave function is zero. Consequently, the subsequent measurement on the second particle P_B will move it to region B^- *regardless of its initial position.* So if P_A is measured first, P_A is deflected up and P_B is deflected down.

Now suppose that P_B is measured first. Since it begins in the top half of the initial wave packet, it will be carried by the $|\downarrow_x\rangle_{P_A}|\uparrow_x\rangle_{P_B}$ component of the wave function as it goes to region B^+ (Fig. 5.10(*b*)). But the two-particle configuration will now be in a region of configuration space where the $|\uparrow_x\rangle_{P_A}|\downarrow_x\rangle_{P_B}$ component of the wave function is zero, so the subsequent measurement of P_A will move it to region A^-. So if P_B is measured first, P_A is deflected down and P_B is deflected up. *So the temporal order of the measurements determines the outcomes.* The problem here is that relativity tells us that if the two measurements are space-like separate (outside each other's light cones), then there will be an inertial frame where the P_A measurement occurs first and another inertial frame where the P_B measurement occurs first. But since the temporal order of the measurements determines the outcomes in Bohm's

---

[13] Note that the distribution postulate allows for the two-particle configuration to be anywhere where there is positive wave-function support.

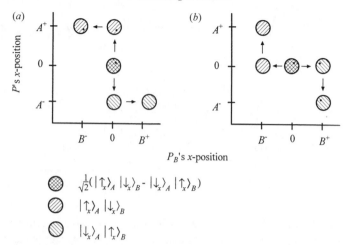

FIG. 5.10 Two EPR experiments in configuration space. (*a*) *A* measures first. (*b*) *B* measures first.

theory, it makes mutually contradictory predictions for the results of the experiment in the two inertial frames. In other words, Bohm's theory is only logically consistent if one chooses a preferred inertial frame. Which is just another way of saying that Bohm's theory is incompatible with relativity.

The conflict between Bohm's theory and relativity is not subtle. If the distribution postulate fails to be satisfied as a boundary condition; that is, if any observer knows more about the position of a particle than the probability density given by the norm-squared of its effective wave function, then that observer can in principle send superluminal signals. Suppose our two EPR observers *A* and *B* agree that *x*-spin up behaviour of *B*'s particle means 'The Yankees won' and *x*-spin down behaviour means 'The Mets won'. Suppose further that observer *B* is in orbit around the Alpha Centauri star system (which looks like just one star, but is actually three) with one EPR particle and suppose that observer *A* is 4.3 light-years away in New York City with the correlated particle. And suppose that observer *A* knows that his particle P_A is closer to A^+ in the wave packet than it is to A^-. The World Series ends (at 12.00 midnight EST) and observer *A* measures the *x*-spin of particle P_A. Since P_A is in the top half of the wave packet, he knows that it will be deflected to region

A^+ regardless of whether his Stern–Gerlach device is set up to deflect the $|\uparrow_x\rangle_{P_A}|\downarrow_x\rangle_{P_B}$ component of the wave function into the region A^+ (call this Setup 1) or set up to deflect the $|\downarrow_x\rangle_{P_A}|\uparrow_x\rangle_{P_B}$ component into region in A^+ (call this Setup 2). That is, observer A knows that if he makes his measurement using Setup 1, then the two-particle configuration will end up associated with the $|\uparrow_x\rangle_{P_A}|\downarrow_x\rangle_{P_B}$ component of the wave function and B's particle will exhibit x-spin down behaviour; and he similarly knows that if he makes his measurement using Setup 2, then the two-particle configuration will end up associated with the $|\downarrow_x\rangle_{P_A}|\uparrow_x\rangle_{P_B}$ component and B's particle will exhibit x-spin up behaviour.

When the Yankees win, A makes his x-spin measurement using Setup 2 (and his particle is deflected up, which in this case means that he got x-spin down). Observer B then, one second after 12.00 midnight EST (in the earth frame), measures the x-spin of his particle; he gets x-spin up, so he almost instantaneously learns that the Yankees won. About four years, four months later, the television signal from the game reaches Alpha Centauri, and observer B and his astronaut colleagues watch it on their portable TV. Observer B wins the pool again; and by dint of his colleagues' firmly held relativistic intuitions, no one suspects cheating.

How seriously one takes this conflict with relativity to be depends on one's physical intuitions. Of course, we have no empirical reason to suppose that superluminal signalling is possible. But in the context of Bohm's theory this just means that we have no empirical reason to suppose that the distribution postulate is false. If the distribution postulate is true, then Bohm's theory makes the same statistical predictions as the standard collapse theory for the results of EPR experiments, and since we have a no-signalling theorem for the standard theory of quantum mechanics, we have a no-signalling theorem for Bohm's theory (with the distribution postulate) as well. Similarly, while there would have to be a preferred frame and while there would in fact be superluminal causal relations, if the distribution postulate is true, then one would never be able to detect the preferred frame. A proponent of Bohm's theory might then argue that the failure of covariance is not so bad since the distribution postulate will ensure that the world always *appears* to be consistent with relativity. This line will not, however, be well received by most physicists, who firmly believe that we have strong theoretical reasons for requiring a Lorentz-covariant dynamics.

Bohm's theory provides the fine-grained causal explanation of the quantum statistics that Einstein seemed to want, and, as he also wanted, quantum probabilities here are ultimately due to our ignorance. But the price for such a theory is a direct conflict with relativity, a price that Einstein was certainly unwilling to pay.[14]

5.5 *Position as the preferred physical property*

Another potential problem for Bohm's theory is the special role played by position in the theory. Since position is usually taken to be the only real physical quantity on Bohm's theory, one can only provide an account of our determinate experiences and beliefs in the theory if one assumes that determinate particle positions (together with the wave function) are sufficient to guarantee that our most immediately accessible measurement records are determinate. But is it in fact true that determinate positions guarantee determinate measurement records? Bell sometimes argued that they would (1987: 128, 166), and Albert argued that they might not (1992: 170–6).

It seems to me that one should worry about taking position as the only determinate physical quantity. Suppose that my brain physiology and the connection between my brain states and mental states is such that my most immediate records of my experience are in terms of *energy*. If this were true, and it might be for all I know about how my brain works, then making positions determinate, as Bohm's theory does, would not necessarily make my most immediately accessible records of my experience determinate. That is, whether Bohm's theory can account for the experiences and beliefs of a particular sentient being depends on the details of how measurements are performed, the being's physiology, and the relationship between physical and mental states. The problem is that while making some physical quantities determinate would do nothing to account for our determinate experiences and beliefs, it is difficult to know whether and to what extent determinate particle positions would provide such an account.

But given the way that a typical system interacts with its environment, it is very difficult to make a record without correlating the position of *something* with the record. Consequently, one would expect that most of our records involve some sort of position correlation. If the position of any

---

[14] Bell always liked Bohm's theory and he thought that it was just the sort of theory that Einstein should have liked. While Bohm's theory is strikingly nonlocal, Bell thought of his own famous no-go theorem not as an argument against Bohm's theory but rather as an argument that one could do no better than Bohm's theory. See Bell's (1987) collection for a history of his thought on Bohm's theory, the implications of Bell's theorem, and his interpretation of Einstein's intuitions.

system in the environment of an observer becomes well-correlated with the observer's brain record (recorded in terms of energy), then there would be a determinate record of the observer's result in terms of the position of that system. And since the observer would then have the sure-fire dispositions of someone with a determinate belief, one might argue that such decoherence effects would in Bohm's theory provide an account of our experience. But is having the same physical dispositions as someone who believes something (in this case the result of a measurement) a sufficient condition for believing the thing? In our discussion of the bare theory we assumed that it was, but whether one really wants to assume this will depend on one's best theory of mind.

The problem here is part of a more general one. Just as there is a matter of fact about what determinate physical properties would explain my determinate mental state (at least at a time), there is presumably a matter of fact about what the real, intrinsic physical quantities are. Consequently, one should probably worry about choosing particle position as the determinate physical quantity unless one can argue (1) that we have good reason to suppose that by making it determinate we can account for our determinate experiences and beliefs and (2) that position is in fact the only real, intrinsic property of particle. But it would be a mistake to argue too strongly for (1) or (2). For one thing, one might worry over the odds that of all the physical properties that might in fact be determinate that position is the one. For another, we have good reason to suppose that particle position is not the physical quantity that we ultimately want determinate. In order to have a Bohm-like quantum field theory, rather than particle positions, one would presumably want to make some field quantity or another determinate, and we would want *this* field quantity to then explain our determinate experience.[15] But if we ultimately want to end up with a quantum field theory, then we presumably do not want to argue too strongly right now that it is determinate particle positions and only determinate particle positions that would explain our determinate experience. My point is simply that since we are not in the position (so to speak) to say which physical quantity we ultimately want to take as always determinate, it would be a mistake to defend Bohm's theory as it stands by arguing for position as the one and only always

---

[15] Note that determinate particle *positions* would not do much to explain the behaviour of neutral K mesons that we discussed in Ch. 1. This is an example of the sort of thing a successful field theory would explain. The problem is that because of the special role that position plays in Bohm's theory (and the associated problem with covariance), it is difficult to use it as the foundation of a relativistic field theory.

determinate physical quantity. Perhaps we ought to drop the commitment to determinate position and settle for a commitment to finding a satisfactory Bohm-*like* theory (one where it may not be position that is determinate).

5.6 *The limiting properties in the context of Bohm's theory*

There is one more thing that I should like to discuss before I consider the many-worlds theories. It is not a problem with Bohm's theory. It is just a piece of trivia concerning the relationship between the bare theory and Bohm's theory that may make each theory a little clearer. It has to do with the bare theory's limiting properties and a subtle difference between the two theories.

Consider an experiment in Bohm's theory where M measures the x-spin of a series of particles in eigenstates of z-spin whose positions are all in fact in the top half of their effective wave packets (Fig. 5.11). And suppose that, because of this last fact, each x-spin measurement is guaranteed to yield the result x-spin up. Suppose M contains two boxes labelled *yes* and *no* and that there is a particle P that starts in the yes-box and moves to the no-box if and only if M ever records the result x-spin down. According to Bohm's theory (just as in the bare theory), M approaches a quantum-mechanical state where all of the quantum probability moves

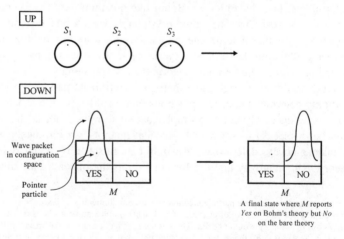

FIG. 5.11 A difference between Bohm's theory and the bare theory.

out of the yes-box and into the no-box. In the bare theory we interpret this limiting state as a state where P is determinately in the no-box and where M thus reports that it recorded at least one x-spin down result. Of course, on the bare theory, M would never determinately record any x-spin results on this experiment (but it none the less ends up with the sure-fire disposition to report that it recorded at least one x-spin down result). This, however, is *not* the limiting disposition of M on Bohm's theory. On Bohm's theory, since all of the measurement results were in fact x-spin up, and since the position of P is determined by the actual positions of the measured particles (the directions that they are in fact deflected in the x-spin measurements), P would simply stay in the yes-box in this experiment, even in the limit, and thus, contrary to the bare theory's prediction, M would have the sure-fire disposition to report (accurately) that it recorded no x-spin down results. More generally, the bare theory predicts that M would approach a state where it reports that half of its results were x-spin up on this experiment, but here, given the unusual initial configuration, Bohm's theory predicts that M would always report, even in the limit, that all of his results were x-spin up.

One reaction to this would be to point out that the initial configuration required for this thought experiment would virtually never obtain. Given the distribution postulate, the probability of all of the object particles initially being in the top half of their effective wave would be zero. There is, however, always at least one question such that the composite system will approach a state where the bare theory predicts that M would answer the question with *no* but where Bohm's theory predicts M will answer the question with *yes*.

Suppose that the initial particle positions are randomly distributed so that half are in the top half and half in the bottom half of their effective wave packets. When M measures the x-spins of the particles, it will always record *some* determinate sequence of results, and will, at least in principle, always be able to report reliably what its results were. But for any particular sequence of results, the composite system will approach a state where M reports on the bare theory that it did not get *that* sequence of results (because the sum of the norm-squared of the coefficients on the terms that describe M as having recorded some other sequence of results goes to one in the limit). Consequently, if we ask M whether it got such-and-such sequence of results, where such-and-such was in fact the sequence that he recorded, then he would answer *yes* on Bohm's theory, but such an observer would always answer *no* on the bare theory.

On Bohm's theory, a particle can have a determinate position when its wave function is not an eigenstate of position. But one can also tell stories (like the one above) where the wave function is in one place and the particle is somewhere else entirely.[16] The moral is just that the eigenvalue–eigenstate link can fail in both directions in Bohm's theory *even for position*.[17]

Neither Bell's Everett (?) theory nor Bohm's theory are very close to Everett's description of his relative-state formulation. They require one to choose an always determinate physical quantity and to provide an auxiliary dynamics for it, and Everett said nothing to suggest that he wanted an always determinate quantity or an auxiliary dynamics. These theories also each violate the principle of state completeness, which was something to which Everett was clearly committed. Finally, Everett explicitly said that his formulation of quantum mechanics was better than Bohm's theory, which allows one to make the rather obvious inference that his formulation was not Bohm's theory (and probably not Bell's Everett (?) theory either). On the other hand, we shall see that when we try to make sense of the usual interpretations of Everett, we often end up with a theory that is very much like one or the other of these two theories.

---

[16] As a very simple example, consider an x-spin measurement by an *ideal* Stern–Gerlach device where the initial wave packet is an eigenstate of z-spin and the particle is precisely in the middle of the wave packet: the two x-spin components will separate, and the particle will be left stranded in a region with no wave function support.

[17] On the topic of exotic states in Bohm's theory, consider what would happen (in a one-particle universe) if the particle started in a position where the wave function was zero and the wave function was initially trapped in an infinite potential when the potential was turned off. The wave function would immediately spread out and sweep the particle out of space entirely—that is, since none of the probability can pass the particle (given the particle dynamics), one can show that for every distance d from its initial position, the particle must be farther away from its initial position than d the instant after the wave function is released. In this sort of exotic situation, then, Bohm's theory exhibits a kind of indeterminism. See Earman (1986) for examples of such indeterminism in classical mechanics.

6

MANY WORLDS

THERE are many many-worlds interpretations.[1] The DeWitt–Graham interpretation, however, has been particularly popular. Indeed, it is probably their splitting-worlds interpretation that most people think of when they think of Everett.

6.1 The splitting-worlds interpretation

In the preface to their 1973 anthology *The Many-Worlds Interpretation of Quantum Mechanics* Bryce DeWitt and Neill Graham say that Everett's interpretation of quantum mechanics

denies the existence of a separate classical realm and asserts that it makes sense to talk about a state vector for the whole universe. This state vector never collapses and hence reality as a whole is rigorously deterministic. This reality, which is described *jointly* by the dynamical variables and the state vector, is not the reality we customarily think of, but is a reality composed of many worlds. By virtue of the temporal development of the dynamical variables the state vector decomposes naturally into orthogonal vectors, reflecting a continual splitting of the universe into a multitude of mutually unobservable but equally real worlds, in each of which every good measurement has yielded a definite result and in most of which the familiar statistical quantum laws hold. (1973, p. v)

---

[1] As an example of the range of theories involved, consider Frank Tipler's many-worlds theory. Tipler claims that most sceptics of the many-worlds interpretation 'have a mistaken idea of what the [many-worlds interpretation] really means' (1986: 204). The problem, he says, is that 'many presentations of the [many-worlds interpretation] have made it appear more counter-intuitive than it really is' (1986: 206). One example of this is when it is claimed that the entire universe is split by a measurement. According to Tipler, 'this is not true. Only the observed/observer system splits; only that restricted portion of the universe acted on by the measurement operator M splits.' Tipler observes that this splitting cannot go on for ever: 'since the information stored in human beings is finite, the set of all possible measurements can split a human being into only a finite number of pieces'. He goes on to estimate that a human being can only be split into about '2 raised to the 10^{26} power' pieces. Such calculations suggest that Tipler takes his talk of people splitting quite seriously. But not only is it difficult to see how this notion of splitting people is supposed to account for our experience; it is difficult to see how a split person, in the sense that Tipler seems to require, could even survive. In any case, this, together with the other theories I consider in the chapter, illustrates the range of many-worlds theories out there.

Both DeWitt and Graham seem to take this talk of many, simultaneously existing, and continually splitting worlds quite seriously. DeWitt explains that

This universe is constantly splitting into a stupendous number of branches, all resulting from the measurement-like interactions between its myriads of components. Moreover, every quantum transition taking place on every star, in every galaxy, in every remote corner of the universe is splitting our local world on earth into myriads of copies of itself. (161)

The entire universe then splits into nearly identical copies of itself whenever a measurement-like interaction or quantum transition occurs anywhere. It is, however, unclear here precisely what in the usual linear evolution of the state is to count as a measurement-like interaction or quantum transition. All we know is that the splitting process is such as to ensure that the measurement records resulting from good measurements are determinate in each world. And it is *this* that will explain our determinate measurement results on the splitting-worlds theory.

While we do not know exactly what counts as a measurement-like interaction, the interaction between a good x-spin measuring device M and its object system S is presumably one. Suppose that S is initially in a z-spin eigenstate and M is in a ready-to-make-a-measurement state. Imagine a very simple universe where the initial global state is

$$|\text{ready}\rangle_M 1/\sqrt{2}(|\uparrow_x\rangle_S + |\downarrow_x\rangle_S). \tag{6.1}$$

This state is supposed to describe a universe containing a single world which itself contains two physical systems M (in state $|\text{ready}\rangle_M$) and S (in state $1/\sqrt{2}(|\uparrow_x\rangle_S + |\downarrow_x\rangle_S)$). After the measurement interaction, the linear dynamics tells us that the global state of the universe will now be

$$1/\sqrt{2}(|\text{up}\rangle_M|\uparrow_x\rangle_S + |\text{down}\rangle_M|\downarrow_x\rangle_S). \tag{6.2}$$

This is an entangled superposition of states corresponding to M recording different x-spin results. Concerning such a state, Graham claims that 'According to Everett, this superposition describes a set of simultaneously existing worlds, one for each element of the superposition. In each world the apparatus has a unique pointer reading, the one described by the corresponding element . . . of the superposition' (1973: 232). The superposition above then describes a universe containing *two* worlds, each containing almost identical copies of the pre-measurement systems M and S. In one world the measuring device, call it M_1, recorded the result *up*, and the object system, call it S_1, is in an x-spin up state; and in the other world the measuring device, call this one M_2, recorded the result *down* and S_2 is in an x-spin down state. That is, the measurement-like interaction between M and S somehow caused the initial world to split into two worlds, and in each of these new worlds there is a determinate result to the x-spin measurement. And it is this last fact that explains the determinateness of our experience on the splitting-worlds interpretation—here an *observer* would be split by the interaction between M and S and one of the resulting observers would see the result *up* in one post-measurement world and other observer would see the result *down* in the other post-measurement world.

Given DeWitt (and Graham's) description, then, the splitting-worlds theory might be summarized as follows:

1. There is a state vector that represents the state of the entire universe.

2. The global state evolves according to the usual deterministic linear dynamics and never collapses (one assumes that there is something like a global Hamiltonian that determines this evolution).

3. The universe (physical reality) consists of many mutually unobservable but equally real worlds.

4. A complete description of physical reality requires one to specify the universal state vector and the dynamical variables.[2]

5. The state vector representing the global state naturally decomposes into orthogonal vectors that represent the states of the various worlds. There is exactly one world corresponding to each term in the preferred decomposition of the *global* state and each term describes the *local* state of the corresponding world.

6. The natural decomposition of the global state vector is one where there is a determinate record (typically different in each world) of the result of every good measurement, and this is what explains our determinate measurement records.

It is important to distinguish here between two types of states: there is a *global* state of the universe given by the universal wave function ψ and there is the *local* state of each of the many worlds that comprise

---

[2] It is unclear what DeWitt and Graham meant by a specification of the dynamical variables; but, since the global state alone is not enough, perhaps it is the specification of dynamical variables together with the global state that is supposed to determine the *local* state of each world. On this reading the specification of dynamical variables would amount to a choice of a specific preferred basis in which to expand to global state. In any case, if one stipulates a particular preferred basis, then, as we will see, one resolves the ambiguity concerning what interactions are supposed to count as measurements.

the universe. The same global state would decompose into different local states for different choices of a basis. In order to account for determinate measurement results, one would want each local state to be one that typically describes observers as having determinate measurement records. Choose one such basis as the determinate-record basis.[3] There is one world for each term in the determinate-record expansion of ψ and each term describes the local state of its corresponding world. As the global state evolves in the usual linear way, what worlds there are and what their local states are change as the determinate-record expansion of ψ changes. At a time each world is associated with a complex-valued amplitude given by the coefficient on the term in the expansion describing the state of that world. This amplitude presumably has something to do with the probability of an observer finding himself in that world, but so far it is unclear exactly how probabilities are supposed to work in the theory. Note that once one chooses a preferred basis, one immediately knows what counts as a measurement-like interaction: a measurement-like interaction is any interaction that leads to more terms in the preferred decomposition of the global state. Putting the pieces together, one might picture the x-spin measurement described above as illustrated in Figure 6.1.

Since DeWitt and Graham present their theory as a reading of Everett, one might naturally wonder how well it meshes with what Everett actually said. Everett himself clearly took principles (1) and (2) above to be features of his formulation of quantum mechanics: he insisted that there was a single universal wave function and that it always evolved in the usual linear way. It is difficult to say, however, to what extent he was committed to principle (3). Nowhere in his published works did he ever claim that the terms representing the branches of the state vector were meant to describe different worlds. He said that each term describes a *branch*, that these branches are all equally real, and that each branch describes a different sequence of experiences for an observer; but he also said that after an observer makes a measurement, there is only one physical observer, which is, strictly speaking, false on the splitting-worlds theory. Also, since Everett always insisted that the wave function by itself provided a complete and accurate description of the physical state, he would probably

---

[3] Just as I did in Ch. 3. Such a basis is also called a *pointer basis* since presumably one way to ensure that all good observers have determinate measurement records would be to ensure that the pointer on each measuring device typically displays a determinate measurement result.

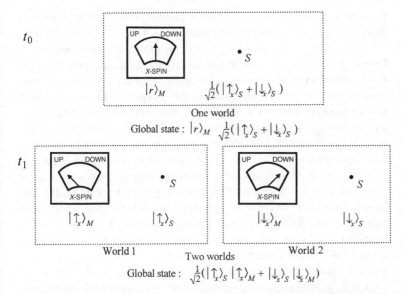

FIG. 6.1 Global states and local states.

object to principle (4) in so far as it requires one to add something to the state description that is not provided by the wave function alone. On the splitting-worlds theory one must choose what physical quantities one wants to be determinate in a world, which amounts to choosing a physically preferred basis. Without a preferred basis, the global state fails to determine the local states. But if anything must be added to ψ to get a physical description, then Everett would presumably disapprove. Also, on a closely related point, while Everett always wrote the state of the observer and object system in the observer's determinate-record basis when explaining why pure wave mechanics makes the same empirical predictions as the standard theory, he nowhere singles out any basis as privileged when he describes his theory; rather, in his discussion of the principle of the relativity of states he goes out of his way to explain that no basis is in anyway privileged. It is unclear exactly what Everett would say about principles (5) and (6), but since one cannot determine the local states in the splitting-worlds theory without knowing the preferred basis, the wave function ψ does not *by itself* determine the complete physical state, which, again, is probably something that Everett would not like.[4]

Failing to agree with what Everett actually said is, of course, not so bad. Indeed, perhaps the most significant virtue of the splitting-worlds theory is that it immediately answers one of the questions that Everett himself continually raised but never clearly answered: it explains the sense in which an observer has a determinate measurement record when the global state is a superposition of incompatible record states. On the splitting-worlds theory, it is because every good observer's measurement record is in fact perfectly determinate in that observer's world that explains each observer's determinate experience. After an x-spin measurement, the observer in one world believes that he recorded the result x-*spin up* because he did. The observer in another world believes that he recorded the result x-*spin down* because he did. And, ideally, both of these observers will inhabit worlds where the local state of the object system will agree with their recorded result. Unlike the bare theory, judgements of the determinateness of measurement results are not to be explained as some sort of illusion; rather, each post-measurement observer really does typically record a single, perfectly ordinary, determinate result. The metaphysics is important here: it is the many-worlds ontology that explains our determinate experience.

While the splitting-worlds theory explains our determinate experience and is in many ways clearer than Everett's formulation of quantum mechanics, it also encounters several problems. Some of these problems are closely related to aspects of the theory that DeWitt and Graham never pin down precisely. I think that some of the problems encountered by the splitting-worlds theory are quite serious, while some of the traditional complaints against the theory are not really problems at all.

6.2 *Traditional and real problems*

6.2.1 *Too many worlds?*

In the preface to their Everett anthology DeWitt and Graham explained that few physicists have taken Everett's interpretation of quantum mechanics seriously.

---

[4] It should also be noted, however, that Everett had a chance to disagree with DeWitt and Graham's reading of his work (when they were putting together their anthology on the many-universes formulation, for example), but, as far as I can tell, he did not tell anyone that they got it wrong.

Looked at one way, Everett's interpretation calls for a return to naive realism and the old fashioned idea that there can be a direct correspondence between formalism and reality. Because physicists have become more sophisticated than this, and above all because the implications of his approach appear to them to be so bizarre, few have taken Everett seriously. (1973, p. v)

Of course, what makes the splitting-worlds interpretation so bizarre is the splitting worlds. DeWitt described his own reaction when he first considered the interpretation:

I still recall vividly the shock I experienced on first encountering this multiworld concept. The idea of 10^{100+} slightly imperfect copies of oneself all constantly splitting into further copies, which ultimately become unrecognizable, is not easy to reconcile with common sense. Here is schizophrenia with a vengeance. (DeWitt and Graham 1973: 161)

One recurrent complaint is that the splitting-worlds theory is ontologically extravagant. Along with its other problems, Bernard d'Espagnat adds that

This theory also has the very unpleasant feature that, in a way, it runs counter to the principle of the economy of assumptions (Occam's razor) which is otherwise known to be so important in science. Indeed it even does so without any restraint, since it goes as far as postulating infinities of completely unobservable worlds or at least *states* of the universe. (D'Espagnat 1971: 445–6)

Presumably the complaint here is not so much that the splitting-worlds theory requires too many theoretical assumptions but rather that it postulates the existence of too many entities. After all, one might argue, if one wants to explain *our* experience, then one only needs to postulate the existence of one world—*our world*. In this sense, claiming that there is more than one physical world would always be postulating entities beyond necessity.

If many-worlds interpretations do not need a plurality of equally real worlds in order to account for *our* experience, then why would one postulate an extravagant many-worlds ontology? Bell thought that there was no good reason (Bell 1987: 97, 133). If one had a successful many-worlds theory, then one could presumably take just one of the worlds to be ours, and simply deny (or not care much about) the existence of the others. One would then be left with the linearly evolving wave function and a parameter that indicates which component of the wave function describes the actual determinate state of our world. In other words, one would be left with a hidden-variable theory. If one took position as the privileged observable, then the particle configuration would select a single term in the position-basis expansion of the global state, the term that describes the actual positions of the particles in the only world there is. As it happens,

Bohm's theory is an example of just such a theory. In such a theory, however, one cannot take the wave function alone as providing a complete
description of the physical state. And here, I think, we arrive at what it
is that a many-worlds proponent believes he is purchasing with his vast
ontology of worlds: he believes that he can take the global wave function
by itself to be a complete and accurate specification of the physical state
of the entire universe. It is not at all clear that this is true—after all, the
global wave function by itself does not even tell us how many worlds there
are, let alone what their local states are; and, even when supplemented
with the choice of a preferred physical quantity, the global state does not
tell us what we should expect to see in *our* world, which one might take
to be a significant descriptive gap. But, even if one does take the global
state to be in some sense complete on a splitting-worlds theory, then one
might wonder whether it is worth the vast ontology of worlds. In any
case, there is a proposed trade-off here between two types of simplicity:
one is purchasing a weak sort of state completeness by postulating a vast
ontology of worlds.

Abner Shimony has also argued against the many-worlds interpretation
on the grounds of ontological extravagance, but for a somewhat different
reason.

It has also been objected that the many-worlds interpretation violated the methodological rule that entities ought not to be multiplied beyond necessity. The only
necessity which has been adduced for maintaining the equal reality of an infinite number of branches is unwillingness to curtail the range of the validity of
standard quantum dynamics. But in view of the fact that the experimental confir-
mation of that dynamics is almost entirely based upon the behavior of microscopic
systems—the only macroscopic evidence being provided by special systems like
superconductors—it is very hazardous indeed to extrapolate the validity of that
dynamics to the universe as a whole. (Shimony 1989: 393)

While I agree that ontological extravagance may purchase little in the
splitting-worlds theory, I do not think that the fault lies with the quantum
dynamics. First, while it might turn out that the linear dynamics does
not correctly describe the time-evolution of all physical systems, it does
seem to describe correctly the time-evolution of every physical system
that we have been able to check by performing appropriate interference
experiments. Also, the linear dynamics is particularly simple and has
other virtues that one might look for in a dynamical law.[5] Consequently,

---

[5] It can, for example, be expressed in a covariant form, which is something that is typically
difficult to do with the auxiliary dynamical laws that have been proposed for quantum
mechanics (like the standard collapse dynamics or Bohm's law for particle motions).

it does not seem crazy, at least not to me, to suppose that it is universally valid, assuming, of course, that one can find a satisfactory account for the determinateness of experience. Secondly, the linear dynamics does not by itself lead one to postulate the existence of many worlds (in Bohm's theory the linear dynamics holds for all physical systems whatsoever and yet there is only one world); rather, it is the claim that the linear dynamics always obtains *and* the claim that the usual quantum-mechanical state provides a complete physical description *and* the desire to account for the determinateness of experience in as familiar a way as possible that together suggests the wild ontology. Thirdly, perhaps more than anything else, Shimony's argument points out a problem with Occam's razor. Very few of the choices made in constructing physical theories are *necessary*. A good methodological principle would admonish one not to postulate entities beyond what seems appropriate given everything one wants from one's best physical theories. How seriously one should worry about the ontological excesses of the splitting-worlds theory depends on what other things one wants from quantum mechanics in addition to a parsimonious ontology and what other theoretical options are available.

An extravagant ontology might, of course, be taken to compound whatever other problems are faced by the splitting-worlds interpretation. Michael Lockwood has argued that the splitting-worlds interpretation 'is a possible view, though I can see no good reason why anyone should wish to adopt it. Essentially, it just piles ontological extravagance on top of all the other difficulties that beset the von Neumann approach' (Lockwood 1989: 225–6). And R.I.G. Hughes has similarly argued that far from placing us in a position better than von Neumann's interpretation, in this interpretation 'any transition not governed by Schrödinger's equation is now accompanied by an ontological outburst' (Hughes 1989: 293–4). I shall discuss why one might take the splitting-worlds theory to share some of the same problems as the standard von Neumann–Dirac theory when I discuss the preferred-basis problem, but even here one can begin to see the problem: DeWitt and Graham's world-splitting rule tells us that worlds split whenever a measurement-like interaction occurs, but they never explain precisely what counts as a measurement-like interaction; rather, this is determined by one's choice of a preferred basis in the theory, which is never made explicit. But note that if one *did know* what it took to count as a measurement-like interaction here, then one would be able to solve the measurement problem in the standard collapse theory by stipulating that the global wave function collapses whenever precisely that sort of interaction occurs.

6.2.2   *The feeling of splitting*

The earliest readers of Everett wondered why we do not notice the constant splitting predicted by the theory. DeWitt's explanation, like Everett's, was that, to the extent that an observer can be regarded as simple automaton, the laws of quantum mechanics are such that they would not allow one to feel oneself split (DeWitt and Graham 1973: 161).

DeWitt's argument goes as follows. First, he assumes that 'If the splitting of the universe is to be unobservable the results [of repeated measurements] had better be the same' (1973: 161). Then he appeals to the argument that Everett himself gave to account for the repeatability of observations that they would in fact be the same (if a second observer *N* perfectly repeated *M*'s measurement, then *N* and *M* would agree on the result of the measurement in both post-measurement worlds). And from this DeWitt concludes that 'The splitting into branches is thus unobserved' (161). But, of course, this conclusion would only follow if one believed the *converse* of DeWitt's initial assumption. That is, one would have to believe that if the results of repeated measurements were the same, then the splitting of the universe would be unobservable; but there is no reason that I can see for believing that the antecedent here constitutes a sufficient condition for the unobservability of splitting.

As we saw in Chapter 3, Everett discussed Wigner's-friend-type interference observations, which, in the language of the splitting-worlds theory, would allow one to detect, or at least to *infer* (depending on what one counts as a *detection*), the presence of other worlds. If one made an *A*-type measurement of Wigner's friend, his measuring device, and the particle he had observed, and if one always got the result $+1$, then this would mean that although the friend might have recorded the determinate result *x-spin up* in *our* world, the global state must be a superposition of determinate *up* and *down* record states, and there must consequently be another world where the observer got *x-spin down* for the measurement. And so *A*-type measurements would allow one in principle to detect the presence of other worlds. But if other worlds can be detected, then the process of splitting is also (at least in principle) detectable. Consequently, if we do not notice the process of splitting, then it must be that we are typically not set up to make the sort of measurement that would detect such a process at work, which, if the preferred basis of the theory is anything like pointer position, is certainly right (again, it would be extremely difficult to perform anything like an *A*-type measurement on a macroscopic system like Wigner's friend).

For his part, DeWitt insisted that 'no experiment can reveal the existence of the "other worlds"' (1973: 165). None the less, he also seems to have noticed that there were observations like an *A*-measurement of Wigner's friend that would show that the local state of one's own world was not complete (164). But these are precisely the same experiments that would allow one to infer the presence of other worlds and determine their local states.[6]

6.2.3 *Compatibility with other physical theories*

Another problem with the splitting-worlds theory concerns whether or not it is compatible with other physical theories and basic conservation principles.

Many critics of the theory have worried about the fact that it is incompatible with the conservation of mass since if the world splits into N copies, there would then be N times as much stuff as there used to be.[7] It does not seem to me, however, that the violation of the conservation of mass itself poses a serious problem for the splitting-worlds theory. Each world resulting from a split would contain the same amount of stuff as the initial world, so it would be easy to explain how we ended up believing in the conservation of mass when mass was in fact not conserved.

There are, however, closely related problems that are serious. Consider, for example, the question of *when* a world splits. According to the splitting-worlds theory, a world splits when a measurement-like interaction occurs, but according to relativity there is no absolute matter of fact concerning when any event occurs—indeed, as I have already discussed, observers in different inertial frames will typically even disagree about the temporal order of events. But when a world splits is presumably a frame-*independent* matter of fact—*it seems to be precisely the sort of event that could not depend on one's inertial frame since there is presumably a simple matter of fact about whether the world is split or not.* Along similar lines, Hughes has complained that the splitting-worlds theory fails to provide a causal explanation for how 'a measurement of spin causes a global bifurcation of space-time' (1989: 293–4). But again if one identifies a world with a space-time manifold, I do not understand how a

---

[6] See Albert (1986).

[7] See Healey (1984: 594), Albert and Loewer (1988: 198–203), and Lockwood (1989: 228–9) for examples of this objection.

world *could* split—*when could it happen* when time might be thought of here as a coordinate property of the manifold itself!? Unless one can find some way to describe the process of a world splitting without choosing a preferred inertial frame, then one has a direct contradiction with relativity. In defence of the splitting-worlds theory, however, one might note that other theories of quantum mechanics, the standard collapse theory and Bohm's theory for example, do not mesh well with relativity. But, then again, this may not provide much consolation.

6.2.4 *What it takes to be a world*

This is a rather different sort of criticism.[8] If one supposes that the state of a world determines the behaviour of physical systems in that world in so far as their behaviour is determined, then the splitting-worlds theory is incompatible with the usual linear dynamics; that is, under a plausible assumption about what it means for one of the worlds in the splitting-worlds theory to be *a world*, that its local state provides the most complete description possible of the dispositions of the physical systems inhabiting it, the splitting-worlds theory is logically inconsistent. How serious one takes this problem to be depends on how strongly one believes that the complete state of a world must provide the most complete description possible of what will happen in that world.

Consider what happens when one makes an A-type measurement in the splitting-worlds theory. Suppose that M performs an x-spin measurement on S and ends up in the global state

$$1/\sqrt{2}(|\text{up}\rangle_M|\uparrow_x\rangle_S + |\text{down}\rangle_M|\downarrow_x\rangle_S). \tag{6.3}$$

Given an appropriate preferred basis, this global state describes two worlds: one with local state $|\text{up}\rangle_M|\uparrow_x\rangle_S$ and the other with local state $|\text{down}\rangle_M|\downarrow_x\rangle_S$. Suppose that these local states determine, in so far as they are determined, all outcomes of measurements in the respective worlds, and suppose that M sets out to measure an observable A which has this state as an eigenstate with eigenvalue $+1$ and every orthogonal state as an eigenstate with eigenvalue -1. If M_1 measures A of $S_1 + M_1$ in the first world, the world where the state of the composite system $M_1 + S_2$ is $|\text{up}\rangle_M|\uparrow_x\rangle_S$, and if this state must provide the most complete possible description of what will happen, then M's result must be a matter of chance where the probability of getting the result $+1$ is $1/2$ (since the

---

[8] See Albert and Barrett (1995).

local state of this world lies at a 45-degree angle to the eigenstate of A with eigenvalue $+1$). And, by a similar argument, one would expect that the measuring device M_2 in the other world would have the probability of $1/2$ of getting the result $+1$ for an A-measurement on the composite system $M_2 + S_2$. But the linear dynamics requires that M have the sure-fire disposition of reporting $+1$ as the A-measurement result. So we have a contradiction here between the linear dynamics and the assumption that the local state of a world completely determines the physical dispositions of systems in that world. Put the other way round, the linear dynamics *requires* that the result of the A-measurement be $+1$ in each world, but there is no way to infer this result from the local state of either world.

Clifton (1996a) gives a particularly nice example that shows that if one accepts the implications of the linear dynamics, then the local states of worlds in the splitting-worlds theory can be *completely irrelevant* to the sure-fire dispositions of the physical systems that inhabit those worlds. Consider the two global states

$$1/\sqrt{2}(|\text{ready}\rangle_N |\text{up}\rangle_M |\uparrow_x\rangle_S + |\text{ready}\rangle_N |\text{down}\rangle_M |\downarrow_x\rangle_S) \tag{6.4}$$

and

$$1/\sqrt{2}(|\text{ready}\rangle_N |\text{up}\rangle_M |\uparrow_x\rangle_S - |\text{ready}\rangle_N |\text{down}\rangle_M |\downarrow_x\rangle_S), \tag{6.5}$$

where N is an A-measuring device ready to measure the composite system $M + S$. On the splitting-worlds theory, both of these global states describe the same two worlds: one with local state $|\text{ready}\rangle_N |\text{up}\rangle_M |\uparrow_x\rangle_S$ and the other with local state $|\text{ready}\rangle_N |\text{down}\rangle_M |\downarrow_x\rangle_S$. But since the two global states are orthogonal, the sure-fire dispositions of N *as predicted by the usual linear dynamics* are radically different for each global state. Even though the local states are the same for each global state, in the first global state, an A-measurement must yield the result $+1$, and in the second global state, it must yield the result -1. That is, the local physical state of the world N inhabits has nothing at all to do with the sure-fire dispositions of N.

One could solve this problem by giving up the assumption that the dispositions of a physical object inhabiting a world are determined as completely as possible by the physical state of that world. But it seems odd to me to claim that the events in a world are determined by something other than the state of that world. One would naturally suppose that if something outside the world was responsible for an event in the world, then that thing was not really outside the world after all. In this sense, worlds in the splitting-worlds theory are not really *worlds*.

There is, however, a more serious problem involving the metaphysics of worlds—there is no clear notion for the identity of a world over time in the splitting-worlds theory, and this makes it difficult to understand what empirical predictions the theory makes.

6.2.5   *Identifying worlds*

Jeremy Butterfield has noted that few many-worlds proponents provide a clear notion of the identity of a world over time. Most authors define branches or worlds at an instant by associating one branch or world with each term in some preferred expansion of the global quantum state, but they typically fail to explain what makes a world the same world at different times. Butterfield argues that this lacuna is often difficult to see because the talk of a branch or the world splitting is 'a metaphor so vivid that we tend to forget that we have been given no account of transtemporal identity for branches' (1996: 6).

This is certainly a problem for the splitting-worlds theory. What we need is a *connection rule* that tells us how to hook together local states at different times into histories. If we had such a rule, then each history would pick out a different world, and we would have a notion of the identity of a world over time. But without such a rule, or some other way to identify the same world at different times, it is difficult to see how one could identify the same observer at different times either. And without a clear notion of the identity of an observer over time, it is senseless to ask whether the theory makes the right empirical predictions for the measurement outcomes of a particular observer. If one has no notion of the identity of an observer over time, then one obviously cannot predict an observer's future experience. In so far as it lacks a notion of the identity of a world over time (and thus no notion of the identity of an observer over time), the splitting-worlds theory is thus empirically incoherent (and no better off than Bell's Everett (?) theory). But if one adds a connection rule to the theory, then this further (because one also needs to choose a preferred basis) detracts from the theory's simplicity.

One cannot judge whether a many-worlds theory is acceptable without first getting the metaphysics straight since the details of the metaphysics are important to whether and how the theory accounts for an observer's determinate records, experiences, and beliefs. For one thing, how we understand the metaphysics affects whether and how we might understand quantum probabilities.

6.2.6 *Whence probability?*

Suppose that an automaton M repeats the same measurement on each of a sequence of systems in the same state S_1, S_2, ... Let $|\psi_n\rangle$ represent the state of M and the first n object systems after n measurements and let $|\psi'_n\rangle$ be the vector that represents the state of M and the first n object systems after n measurements *but without the terms that describe sequences of results that are nonrandom or have relative frequencies that differ from those predicted by the standard formulation of quantum mechanics.* As we saw earlier, the difference between $|\psi_n\rangle$ and $|\psi'_n\rangle$ goes to zero in the limit as n gets large. From this fact DeWitt concludes that in the context of the splitting-worlds theory (1) 'Each automaton in the superposition sees the world obeying the familiar statistical quantum laws' and (2) 'The conventional probability interpretation of quantum mechanics thus emerges from the formalism itself' (DeWitt and Graham 1973: 163). Such conclusions are, however, at best misleading. On the splitting-worlds theory it is typically false that each world, or even most worlds in the usual sense of *most*, exhibit the familiar quantum statistics. Indeed, the vast majority of worlds, in the usual counting sense of most, will obey laws very different from the familiar statistical quantum laws. But perhaps more importantly, the splitting-worlds theory, as it stands, provides nothing like the conventional probabilities of quantum mechanics. Something is missing from the splitting-worlds theory. Exactly what one takes it to be depends on exactly what one wants the theory to explain.

David Albert and Barry Loewer argued that

Whatever merits DeWitt's argument for his claim that the probability interpretation emerges from the quantum mechanical formalism may have, it doesn't address the really difficult problem that the dynamical equations of motion are deterministic. Since, according to the [splitting-worlds theory], it is certain that all outcomes of the measurement will occur and be observed ... what can be meant by saying that the probability of a particular outcome $= c^2$? If probability is to be introduced into the picture, it must necessarily be by *adding* something to the theory.

So what needs to be added?

... we might say that some of the worlds ... are more 'actual' than others ... Then probability can be identified with the probability that the actual world will follow a particular branch ... The trouble with this suggestion is not only that it is mysterious (what distinguishes the more actual worlds from the less actual ones?), but also that it gives up the central feature of the SWV [splitting-worlds view], that the state-function entirely exhausts what there is to be said about the physical world. (Albert and Loewer 1988: 201; see also Albert 1992: 114–15)

Albert and Loewer suggest that we need some way of marking *exactly one* future world as actual.

Butterfield agrees that something is missing, but he disagrees with Albert and Loewer concerning what it is. Rather than there being a problem with taking all worlds to be equally actual, Butterfield takes the problem to be that there is no account for the persistence of branches: 'I think that Everettians may well be able to explain why the quantum algorithm's numbers deserve the name "physical probabilities". But doing so depends on filling a lacuna: giving an account of the persistence for branches' (1996: 14).[9]

It seems to me that there is a sense in which both Albert and Loewer and Butterfield are right. If one followed Albert and Loewer's suggestion and added to the theory a parameter that marked exactly one Everett branch as actual, then one would end up with a hidden-variable theory in which there would be no special problem about making sense of probabilities. On the other hand, if we had a way of identifying worlds and observers over time—what I understand Butterfield to mean by 'an account of the persistence of branches'—one might be able to say that all worlds are equally actual and still make sense of probabilities. Further, note that if one followed Albert and Loewer's suggestion of marking one world as actual, then one would automatically have a way of identifying worlds and observers over time since there would only be one world. One might take their suggestion then to provide a sufficient but not necessary condition for making sense of quantum probabilities in the context of a many-worlds theory.

In order to say that one has captured the conventional quantum probabilities, one must be able to assign probabilities to measurement outcomes— after all, the standard collapse formulation of quantum mechanics allows one to do just this. Suppose I perform an x-spin measurement on a system S initially in the state

$$1/2|\uparrow_x\rangle_S + \sqrt{3}/2|\downarrow_x\rangle_S. \tag{6.6}$$

The resultant state will be something like

$$1/2|\uparrow_x\rangle_M|\uparrow_x\rangle_S + \sqrt{3}/2|\downarrow_x\rangle_M|\downarrow_x\rangle_S. \tag{6.7}$$

[9] That every world is actual does not necessarily mean that every world is *my* world. If one assumes that it makes sense to identify one world as *my* world (or *our* world), then one might talk about the (epistemic) probability of my (our) world having such-and-such properties.

In order to capture the usual predictions of quantum mechanics, one would like to say that the probability that I would end up in the world described by the first term is $1/4$ and that the probability that I will end up in the world described by the second term is $3/4$. If one could say this, then one would have an explanation for the fact that I get results that tend to be close to the usual statistical predictions of quantum mechanics. But, as Albert and Loewer point out, on the splitting-worlds theory, as it stands, one cannot say this. Rather, the splitting-worlds theory tells me that *both* worlds are equally real and that there will be a fully real copy of me in each world, so presumably neither has a better claim than the other to being the one that *I* end up experiencing. And here one can see Butterfield's point too. The problem is at least in part one of having no clear notion of the identity of the observer over time—one cannot assign probabilities to *my* future experience because there is no clear notion of which future observer is *me*.

Suppose we have a machine where whenever an observer walks into one side of the machine wearing a blank T-shirt, two perfect copies of the observer walk out the other side of the machine, but one with a T-shirt marked 1 and the other with a T-shirt marked 2. One might imagine this machine as something like a (malfunctioning) transporter from *Star Trek*: it reads off the information it needs to construct a copy of the observer, destroys the observer in the process, then uses the information to construct *two* copies (rather than the usual single copy constructed by a functioning transporter). Imagine walking into such a machine. What is the probability that you would emerge as the observer with a 1 on his T-shirt? One might suppose, by symmetry, that the answer is $1/2$, but it seems to me that this cannot be right. Rather, it seems to me that, by symmetry, there can be no good reason to identify either one of these observers as the *real* me. Consequently, it seems to me to be senseless even to ask what the probability is of *me* ending up as the observer with the T-shirt marked 1. More generally, to talk about the probability of a particular object having some property in the future, one must be able to identify *that object* in the future. At least part of the problem in getting the usual probabilistic predictions from the splitting-worlds theory then arises because it provides no way to identify a particular observer in the future.

The standard collapse formulation of quantum mechanics (together with the rules-of-thumb we use for determining what constitutes a measurement interaction) makes very accurate statistical predictions concerning what we should expect to see when we perform experiments. Indeed, it is the fact that it so successfully predicts what we should expect to see that has made it such a useful theory. Unless one can provide similarly accurate statistical predictions concerning our future experience, one cannot claim to have captured the conventional probability interpretation of probability. The problem is not that the splitting-worlds theory fails to make the right empirical predictions; rather, the problem is that because the splitting-worlds theory does not provide transtemporal identities, it does not make *any* empirical predictions concerning *my* future experience.

One might try adopting a weaker standard of empirical adequacy and argue that rather than capturing the standard predictions of quantum mechanics concerning future events, all we really need is for a theory to explain why a *typical* observer should expect to record the usual quantum statistics *at a time*. If we can do this, then, the argument might go, we have 'deduced' the usual quantum probabilities. It seems to me that such a theory would make *much weaker* empirical predictions than the standard theory. But further, I do not think that the splitting-worlds theory as it stands can accomplish even this.

Consider trying to use the splitting-worlds theory to make bets about where the Eiffel Tower is *right now*. Suppose that the universal wave function assigns a small but nonzero amplitude to me having all the beliefs that I have as I write this and the Eiffel Tower being in Pittsburgh (as it presumably would given the usual dynamics). So I bet a friend $10 that the Eiffel Tower is in Pittsburgh right now. This is presumably a bad bet, *but why*? The splitting-worlds theory doesn't say. There are worlds where I believe what I believe and the Eiffel Tower is in Pittsburgh, and in those worlds I would win the bet. There are worlds where I believe what I believe and the Eiffel Tower is not in Pittsburgh, and in those worlds I would lose the bet. But there is nothing in the splitting-worlds theory as it stands that tells me which sort of world I inhabit right now or even which sort I should *expect* to inhabit right now. More generally, there will be some worlds where my records exhibit the usual quantum statistics and some where they do not, and the splitting-worlds theory gives me no reason to expect that I inhabit the former sort of world right now.

As it stands, then, the splitting-worlds theory does not even explain why one should expect to record the usual quantum statistics *right now*. The best that it can do is to tell us that some observers in some worlds record the usual quantum statistics while other observers in other worlds do not. The right conclusion to DeWitt's argument above is that most worlds *in the norm-squared measure* exhibit the usual quantum statistics, but this fact does not by itself tell us anything about what we should expect to be true in our world even at a time. Part of the problem with getting the standard statistical predictions has to do with the lack of a notion of transtemporal identity for observers, but another part of the problem is that there is nothing whatsoever in the splitting-worlds theory to tie *our* experience (even at a time) to the fact that most worlds in the norm-squared measure exhibit the usual quantum statistics.

DeWitt seems to have noticed that the splitting-worlds theory alone did not explain why one should expect the usual quantum statistics. After presenting his interpretation of probability, he confessed that

The alert reader may now object that the above argument in circular, that in order to derive the *physical* probability interpretation of quantum mechanics, based on sequences of observations, we have introduced a *nonphysical* probability concept, namely that of the measure of a subspace in Hilbert space. (DeWitt and Graham 1973: 163)

And he continued:

It should be stressed that no element of the superposition is, in the end, excluded. All the worlds are there, even those in which everything goes wrong and all the statistical laws break down. (186; see also 163)

DeWitt wanted to find some way of explaining why one should none the less expect to find oneself in a world that exhibits the usual quantum statistics. His main argument was anthropic: perhaps life simply fails to evolve in maverick worlds where the usual quantum statistics fail to hold, 'so no intelligent automata are around to be amazed by it' (186; see also 163). But even if one finds an anthropic argument appropriate here, it is difficult to see why one should take the usual quantum statistics as a precondition for the possibility of sentient life. Presumably, there are worlds where life evolved very much as we believe it did in ours but where observers none the less record very different statistics from those predicted by the standard theory. In the event that one does not find this anthropic argument convincing, DeWitt also argued that 'It is possible that maverick worlds are simply absent from the superposition' (163). If this were right, then it would certainly explain why we should expect to get usual quantum statistics in our world (since *every* world would exhibit the usual quantum statistics), but in order to get what DeWitt seems to want here, one must abandon the linear dynamics, Everett's model of measurement, or change the splitting-worlds theory in some other fundamental way, and DeWitt offers no concrete suggestions. Further, to fix the theory so that the only terms in the global state are terms where the usual quantum statistics are exhibited strikes me as being manifestly *ad hoc*. It seems to me that such desperate suggestions show that DeWitt himself felt that something important was missing from the splitting-worlds account of probability.

DeWitt did not want to add anything to the global state that would pick out one world as actual; rather, he held with Everett that all branches of the superposition are 'equally real'. But he also claimed that 'Each branch corresponds to a possible universe-as-we-actually-see-it' (DeWitt and Graham 1973: 163). It is unclear from what DeWitt said precisely how this was supposed to work, but if the splitting-worlds theory could somehow make predictions about what one should expect to be true of the world-as-*we*-actually-see-it, then this would allow for the possibility of explaining why one should expect the world-as-we-actually-see-it to exhibit the usual quantum statistics. This would provide the weaker sort of account of quantum probabilities—the sort that does not require the theory to make predictions for an observer's future experience and thus does not require a notion of the transtemporal identity of observers, but only requires one to account for why one should expect *right now* to have records in this world that exhibit the usual quantum statistics. The explanation might go something like this: (1) one should expect the world-as-we-actually-see-it to be typical, (2) typical worlds exhibit the usual quantum statistics, (3) therefore, one should expect the world-as-we-actually-see-it to exhibit the usual quantum statistics.

But why and in what sense should we expect our world to be *typical*? One answer, not a very good one, is that if there are N alternatives, then one should always expect each alternative to occur with probability $1/N$. This is sometimes called the principle of indifference. If one adopted this principle, then all one would need to show is that most worlds, in the usual counting sense of most, exhibit the usual quantum statistics. It would then follow that one should expect *our* world to exhibit the usual quantum statistics. But there are at least two problems with this. First, the principle of indifference is generally a bad principle to adopt. And secondly, as mentioned earlier, it is simply false that most Everett worlds, in the usual counting sense of most, will exhibit the usual quantum statistics.

It was this second problem that bothered Graham. Graham believed that 'Provided one is ready to accept the existence of multiple, simultaneously existing worlds, Everett's interpretation satisfactorily explains the apparent reduction of the state vector characteristic of quantum measurements' (Dewitt and Graham 1973: 232). But, because one would typically fail to get the right quantum statistics in *most* worlds, he argued that Everett's account of why one should expect to get the usual quantum statistics was inadequate (note that he is reading Everett as a many-worlds theorist). That is, what worried Graham was that (if one considers measurements of observables with discrete spectra) there would typically be fewer worlds where observers see statistics close to the usual quantum statistics than worlds where they see statistics very far from the usual quantum statistics—indeed, in the limit as the number of such measurements gets large, the numerical proportion of worlds where the observers get within ϵ of the right quantum statistics typically goes to zero.

Consider the situation where M measures the x-spin of a sequence of systems, each in the initial state

$$1/2|\uparrow_x\rangle + \sqrt{3}/2|\downarrow_x\rangle. \tag{6.8}$$

The number of terms in the state of M and the first n object systems, when written in the determinate-record basis, will be 2^n after n measurements, and every possible sequence of measurement records will be represented exactly once. It is this last fact that causes the problem. The proportion of length-n binary sequences where about $1/4$ of the elements are $\uparrow_x$ and about $3/4$ are $\downarrow_x$ goes to zero as n gets large. It is only true that most worlds exhibit the right quantum statistics if each term in the initial state here had the coefficient $1/\sqrt{2}$.

Regardless of what the coefficients are in an experiment like that above, the number of length-n binary sequences is 2^n, so the number of terms in the determinate-record expansion of the state of the composite system will be 2^n after n measurements (and there will be exactly one term for each possible sequence of results). The number of length-n binary sequences with $p \uparrow_x$ results will be $\binom{n}{p} = n!/p!(n-p)!$. And the proportion of length-n sequences where the relative frequency of $\uparrow_x$-results is r will be $\binom{n}{rn}/2^n$. What happens is that, as n gets large, almost all of the length-n sequences of results will have about the same number of $\uparrow$-results as $\downarrow$-results (that is, as n gets large, the proportion of sequences where the ratio of $\uparrow$-results to n is within ϵ of $1/2$ approaches 1). The numerical proportion of length-n sequences where the relative frequency of $\uparrow$ results is within 0.1 of $1/2$, for example, is 0.656 for $n = 10$, 0.737 for $n = 20$, 0.846 for $n = 50$, and 0.965 for $n = 100$. But this means that the numerical proportion of length-n sequences where the relative frequency of $\uparrow$-results is close to anything other than $1/2$ goes to zero as n gets large. Which means that if we associate one world with each term in the determinate-record expansion of the wave function (where there is one term for each possible sequence of x-spin results), then the relative frequency of $\uparrow$-results will be $1/4$ in almost none of the worlds in the experiment described above, which means that almost *none* of the worlds will exhibit the right quantum statistics for this sequence of experiments.

The moral is simply that most worlds, in the usual counting sense, will typically exhibit the *wrong* relative frequencies, and in the limit as the number of measurements gets large almost all worlds will exhibit the wrong relative frequencies. In the norm-squared measure, however, Everett's measure, the *measure* of worlds where observers get the right quantum statistics will tend to increase over time, and in the limit, almost all (measure one) of the worlds will have this property (this is a consequence of the general limiting property I discussed in Chapter 4). But Graham believes that Everett does not sufficiently motivate his choice of measure.

In short, we criticize Everett's interpretation on the grounds of insufficient motivation. Everett gives no connection between his measure and the actual operations involved in determining a relative frequency, no way in which the value of his measure can actually influence the reading of, say, a particle counter. Furthermore, it is extremely difficult to see what significance such a measure can have when its implications are completely contradicted by a simple count of the worlds involved, worlds that Everett's own work assures us must be on the same footing. (DeWitt and Graham 1973: 236)

That is, Everett's argument (in the context of the splitting-worlds theory) was that most worlds in the norm-squared measure would have the right quantum statistics, but he did not explain why an observer should expect his world to be typical in this sense (the sense of typical provided by *this particular measure*). What Graham wanted was that the usual quantum statistics be true in most worlds, in the usual counting sense of most; then a typical observer (in the standard counting sense of typical) would get the usual quantum statistics.

Graham's solution was to formulate a new model of measurement in the splitting-worlds theory—a model that predicted that one would measure the usual quantum statistics in *most* worlds.[10] Then, Graham argued,

If we assume our own world to be a 'typical' one [now in the simple numerical sense], then we may expect a human or mechanical observer to perceive relative

---

[10] Graham modelled the usual Everett sequence of measurements as a 'two step' process: first a macroscopic apparatus measures the relative-frequency operator on a collection of identically prepared systems, then an observer reads the pointer on the measuring apparatus, which causes the state of the observer and the apparatus to split into Everett worlds. Graham then argues that given the properties one would expect a real macroscopic measuring device to have, the observed pointer reading will be near the right value for the relative frequency in the numerical majority of Everett worlds (DeWitt and Graham 1973: 238–50): 'We thus conclude that the values of relative frequency near [those values predicted by the standard theory] will be found in the majority of Everett worlds of the apparatus and the observer' (252).

frequencies in accordance with the Born interpretation. Why we should be able to assume our own world to be typical is, of course, itself an interesting question, but one that is beyond the scope of this paper. (DeWitt and Graham 1973: 252)

So even if one *can* get the numerical majority of worlds to exhibit the usual quantum statistics by changing how one models measurements, the theory still fails to explain why we should expect our world to exhibit the usual quantum statistics. As Graham noted, there is still something missing.

Graham was worried that there was nothing in the splitting-worlds theory that tied Everett's norm-squared measure to our experience. This is a good thing to worry about, but why he thought that we would be better off having most worlds *in the usual counting sense* exhibit the usual quantum statistics is mysterious, *unless he accepted something like the principle of indifference*. If one could show that most worlds exhibit the usual quantum statistics, and if one then assumed that each world had an equal probability of being *ours*, then one would have a straightforward explanation for why one should expect our world to exhibit the usual quantum statistics; but what is the justification for this second assumption?

The principle of indifference is not a very good principle of reason. For one thing, if any such principle were in fact true, then it would presumably be very difficult to say when and how to apply it. Since the Eiffel Tower either is in Pittsburgh or is not in Pittsburgh, the principle of indifference seems to tells us, in so far as it tells us anything at all, that the probability of the Eiffel Tower being in Pittsburgh is $1/2$! With this sort of reasoning, one is doomed to failure. But if this is a misapplication of a generally valid principle, then what exactly is the principle and how was it misapplied?

I take the moral to be that we cannot count on an appeal to any universal principle of reason to fill the gap in Graham's explanation for why one should expect to inhabit a world that currently exhibits the usual quantum statistics—that is, there is no reason to suppose that just because there are N worlds, the probability of our world being a particular one of these is $1/N$. Of course, one could stipulate *as a new axiom* of Graham's theory that this is in fact the right probability; then the theory could explain why one should expect to see the usual quantum statistics at a time. But without accepting something like the principle of indifference, it seems to me that there can be no reason to prefer Graham's counting notion of what it means for a world to be *typical* over Everett's norm-squared notion.

Without a universal principle of reason to tell us what we should expect (how we should assign probabilities to various possible alternative worlds being the-world-as-we-see-it), we must rely on the theory itself to explain, if it can, why one should expect to get the usual quantum statistics. Given this, perhaps the simplest thing to do in the context of the splitting-worlds theory would be to stipulate *as a new axiom of the theory* that the probability that a particular Everett world is the world-as-we-actually-see-it at a time is given by Everett's norm-squared measure.

If one can make sense of the notion of the world-as-*we*-actually-see-it, then this revised splitting worlds theory would explain why one should expect the world-as-we-actually-see-it to be one that exhibits the usual quantum statistics: it would be highly probable (at every time) that the world-as-we-actually-see-it is one that exhibits the usual quantum statistics (in our records). But there are still at least two problems: (1) the assumption that one must add to the theory is quite strong—presumably much too strong for one to say that the explanation why one should expect to see the standard predictions of quantum mechanics emerges from the formalism alone and (2) this approach (of trying to make sense of what it means to talk about the world-as-we-see-it, then adding a new axiom to the theory that assigns a high probability to the world-as-we-actually-see-it being one that exhibits the right quantum statistics), if it succeeded, would still fail to predict our *future* experience (because we still do not have transtemporal identity). In this sense, one still fails to get the conventional quantum probabilities from the splitting-worlds theory. Indeed, there is a pretty good argument that nothing like the splitting-worlds theory (a theory where observers keep splitting into almost identical copies) could ever accomplish this. Just as I cannot say anything sensible about the probability that I will be the person wearing the number 1 T-shirt after walking into a person duplicator, I cannot say anything sensible about the probability that I will be the observer that gets *up* when I measure the x-spin of a system in an eigenstate of z-spin on any theory that predicts that there will be two post-measurement observers each with an equal claim to be the future me. Consequently, if one wants a theory that can predict and provide rational expectations for future events, then it seems to me that this perfectly symmetric splitting of observers has to go. And we will consider a way of doing just this shortly.

Exactly what one takes the probability problem to be depends on what one wants the splitting-worlds theory to explain, but the general moral is simple: unless knowing that most worlds in the norm-squared measure will exhibit the usual quantum statistics at an instant explains everything one wants explained about quantum probability, then there is something missing from the account of quantum probabilities given by the splitting-worlds theory as it stands. And again, it seems that no theory in which constantly splitting observers are cloned can make the same sort of predictions about future experience that are made by the standard theory of quantum mechanics because of the transtemporal identity problem. Without a way of identifying observers over time there can be no explanation why *I* should expect to see my future measurement results exhibit the usual quantum statistics. And this means that at the end of the day one cannot recapture the usual quantum probabilities.[11]

6.2.7 *The preferred-basis problem*

The most straightforward version of the splitting-worlds theory, the one that we have been considering here, postulates the existence of a world corresponding to each element in a linear expansion of the universal wave function. But since there are many possible expansions of the wave function, one must choose a preferred basis, a basis that, as a matter of physical fact, determines what worlds there are and their local states given the global state. By choosing a preferred basis one determines what physical quantity is always determinate. As we began to see in the last chapter, however, it is not so easy to choose a preferred basis.

Part of the problem is that it is unclear *which* basis one should choose. Since we want each observer in each world to have determinate perceptions and memories, we want to choose a basis whose elements each correspond to a state where every sentient observer has determinate perceptions and memories, but what basis is that? The basis one should choose in order to make our most immediately accessible measurement records determinate presumably depends on the details of how we conduct our experiments, our physiology, and the relationship between physical and mental states, and we simply do not know enough of the details to make the choice. Indeed, brains are sufficiently complicated that we may never know precisely what physical property would have to be determinate in order for even a single memory to be determinate in a particular brain.

The other part of the preferred-basis problem is that even if we did somehow figure out what determinate physical property would guarantee determinate mental states and then chose that property as the one and only

---

[11] Much work has been done by philosophers on the ways of identity, and one could no doubt formulate a fancy theory of personal identity for the splitting-worlds view (one could take the pre-split observer to be a past temporal part of each post-split observer, etc.). But one would need a fancy way of understanding probabilities to go along with the theory of personal identity—so fancy that it would be difficult (for me at least) to understand the relevance of such probabilities to *my* experience.

always determinate physical property, our theory would end up looking blatantly *ad hoc*. Why should one expect that *exactly that brain property that sentient beings use to record their perceptions* is the only physical property out of a continuously infinite number of properties that one can define that is always determinate? This is really crazy: our most basic physical laws should not depend on details of our brain physiology.

Of course, one might try to argue that this works the other way round: that rather than physical laws depending on our brain physiology, our brain physiology evolved to exploit whatever property was in fact determinate as a matter of physical law. But it is not at all clear how such an evolutionary story would go since there would typically be no evolutionary pressure for an organism to record its memories in the always determinate quantity.

It is important to get clear about this. Consider an observer Jack who records the result of an x-spin measurement in the x-spin of a single particle in his brain (he just correlates the x-spin components of the two wave functions) without correlating the position of anything with either x-spin.[12] The problem is that Jack lives in a world described by Bohm's theory, where only positions are always determinate; consequently, he does not record determinate measurement results, and hence cannot be said to have any determinate beliefs about the outcomes of his measurements. He will none the less *act* just as if he always has perfectly determinate memories of measurement results just like everyone else. After making an x-spin measurement, Jack will have the sure-fire disposition to report that he recorded a determinate result for the same reason that a bare theory observer would; and, just as on the bare theory, this report would be false, because there would be no determinate record of x-spin up and there would be no determinate record of x-spin down (since Jack recorded the result in a contextual property, what the record *is* depends on precisely how Jack reads it; see Albert 1992: 170–6). But if we ask him what his memories are and if he responds by moving his mouth, say, then this will correlate the position of something with his brain state, which on Bohm's theory will endow him with an effectively determinate result, namely the result that he reports. That is, even though Jack in fact recorded no determinate result, as soon as he is called upon to act on his 'memories' in any way that correlated the position of anything with his brain state, he will end up with an effectively determinate memory, and from then on he will act as if this had been his actual mental state all

---

[12] Jack is a close cousin to Albert's John and John-2 (1992: 170–6).

along.[13] And since on Bohm's theory the probability of ending up with a particular effective memory is equal to the probability of ending up with an ordinary memory if one had recorded the result in terms of position in the first place, Jack will get along just as well as everyone else, and there would be no clear evolutionary reason for Jack-type organisms *who typically do not have determinate beliefs about their experience* (at least not until they are required to act on their beliefs) to be selected against. It is the possibility of successful organisms who do not in fact have the sort of determinate experiences that they claim to have that makes it difficult to argue that evolutionary processes would select for a brain physiology such that a sentient observer always records experiences in terms of whatever physical property is in fact always determinate.

The preferred-basis problem, then, is that we do not know what physical property we need to make determinate to account for experience, and even if we did, stipulating that such a *just-right* property was the only determinate physical property would make our theory look *ad hoc*. And finally there is no reason to believe that sentient beings would naturally evolve so that their most immediately accessible records were in terms of the physical property that was in fact always determinate.

It would be nice if there were some way to make the choice of the preferred basis independent of contingent facts about such things as brain physiology and the relationship between mental and physical states, but if our goal is to account for our determinate experience, then it seems that such facts are ultimately relevant, and one can always ask of any proposed preferred basis whether making that particular quantity determinate provides a theory that accounts for our determinate experiences. The inevitable abyss of worrying about the details of brains and such is what makes the preferred-basis problem so annoying. One remedy would be not to require that physical theories account for our experience (as Pauli suggested; see Section 2.5), but then how would we judge their empirical merits? Or one might go the positivist route and try to formulate one's best physical theories so that they predict appearances *directly* without one having to know anything about how brains function. In any case, one cannot help thinking that we must somehow be asking for far too much from our physical theories when we ask them to make precisely the right brain properties determinate at precisely the right times (as one would

---

[13] I am not sure that one would want to say even now that the observer has a determinate *mental* state, but this would depend on the details of one's theory of mind.

want to do by choosing a just-right preferred basis or as Wigner's theory does by stipulating a very special collapse rule).

It seems to me that the preferred-basis problem is as serious a problem for the splitting-worlds theory as the original measurement problem was for the standard theory—indeed, as suggested earlier, one might argue that the preferred-basis problem is just the original measurement problem in another guise. Just as the collapse on measurement is what makes outcomes determinate in the standard theory, the splitting of worlds on measurement is what makes outcomes determinate in the splitting-worlds theory. And since, in the context of the splitting-worlds theory, it is the preferred basis that determines what physical correlations will cause worlds to split, the preferred basis determines what correlations will lead to determinate measurement results. Since choosing a preferred basis in the splitting-worlds theory amounts to choosing what interactions will constitute measurements with determinate results, and since the measurement problem in the standard theory was precisely the difficulty of determining what interactions one should take as generating determinate measurement results (what interactions one should take to cause collapses), the splitting-worlds theory is arguably no better off than the standard collapse theory.

So the analogy between the original measurement problem and the preferred-basis problem is this. The standard formulation of quantum mechanics tells us that a collapse occurs whenever a measurement is made, but it does not tell us what constitutes a measurement. Similarly, the splitting-worlds theory tells us that a world splits (in so far as one can identify the world) whenever a measurement-like interaction occurs in that world, that is, whenever the preferred basis requires the world to split, but it does not tell us which basis is preferred. If the standard theory is to explain the determinateness of experience, then the collapse must occur in such a way that the brain of an observer typically ends up recording a determinate result. Similarly, if the splitting-worlds theory is to explain the determinateness of experience, then worlds must split in such a way that brains typically end up recording determinate results in each world. And not just any preferred basis will do this. The problem is that in each theory one must make a choice (when collapses occur in one and when splits occur in the other), but this choice is constrained by the fact that in the end we want an account of our determinate of experience.

Bell was one of the first to argue that because of the preferred-basis problem Everett's theory was no better off than the standard collapse theory. According to Bell, Everett 'seems to envisage the world as a multiplicity of "branch" worlds, one corresponding to each term ... in the expansion [of the state in some basis]'. In order to do this, however, Bell believed that Everett needed a preferred basis. Further, Bell argued that in order to account for appearances Everett's preferred basis needed to make instrument readings determinate in each branch:

This preference for a particular set of operators is not dictated by the mathematical structure of the wave function ... It is just added (only tacitly by Everett, and only if I have not misunderstood) to make the model reflect human experience. The existence of such a preferred set of variables is one of the elements in the close correspondence between Everett's theory and de Broglie's—where the positions of particles have a particular role.

The problem then becomes one of trying to specify what it takes for something to count as a determinate instrument reading, which Bell took to be fully analogous to trying to specify what counts as a measurement in the standard theory. One of the arguments that Bell gives in favour of Bohm's theory is that, unlike Everett's theory, it explicitly tells us what physical quantity is preferred (Bell 1987: 95–7). But, of course, this is a real virtue only if position is in fact the *right* quantity to choose as always determinate.

In order to argue that Bohm's theory successfully accounts for our determinate experiences and beliefs, one would have to argue that making positions always determinate provides determinate experiences and beliefs; but if one had such an argument, then one could solve the preferred-basis problem for the splitting-worlds theory—choose position as the preferred basis. The problem, of course, is that we have no such argument. On the other hand, it would be great if we knew that determinate *positions* would account for our determinate experiences and beliefs since choosing a well-entrenched property like position as always determinate would look much less *ad hoc* than choosing some special brain property. But again, in so far as we have good reason to want a quantum field theory, we have good reason to suppose that position is *not* the quantity that we want to choose as always determinate. Moreover, even if a particular choice of a determinate quantity did provide determinate records, experiences, and beliefs for *us*, it would *always* be possible to consider deviant observers, like Jack, who would typically fail to have any determinate belief concerning what they observed immediately following an observation.

The preferred-basis problem is often described as a violation of a supposed democracy of bases in the mathematical formalism of quantum mechanics. In response to this complaint, Daumer *et al.* (1996), among others, have argued (in the context of trying to justify the special status of position in Bohm's theory) that it is naive to believe that all Hermitian operators in the Hilbert-space formalism have equal claim to corresponding to physically real, measurable quantities.[14] This is clearly right. One should not find arguments against the existence of a preferred basis from the assumption that all Hermitian operators are created ontologically equal very compelling. On the other hand, I do not think that this point addresses the real preferred-basis problem.

The real preferred-basis problem is not that having a preferred basis violates some basic principle of the ontological democracy of bases but rather that presumably only a very special basis will provide a satisfactory account of our determinate experience, and we do not know which it is; and, if we choose a physically preferred basis solely on the basis of what we believe would make our experience determinate given our physiology and practices, then this choice will look *ad hoc*. There is nothing inherently wrong with choosing a preferred physical quantity; but if one wants the choice to look natural, then one needs to have good *physical* reasons for that particular choice, not reasons having to do with human physiology and practice.

But again, one cannot help thinking that we have made the job of physics too difficult: after all, classical mechanics was never required to provide a detailed account of the determinateness of our experience. There is, however, a significant difference between quantum and classical mechanics. In classical mechanics there is a collection of interrelated physical properties that are always determinate and the collection is rich enough that we tacitly came to believe that these determinate properties (in so far as they evolved in the right way) would ultimately allow us to account for our actual determinate experience.[15] In quantum mechanics, however, the Kochen–Specker theorem (Kochen and Specker 1967) provides very good reason for supposing that we cannot get the right empirical predictions and make all classical properties of a system determinate without sacrificing functional relationships between the properties (in which case one might argue that they are not really the properties we wanted to make determinate). Indeed, it seems that the set of properties that can be simultaneously determinate without sacrificing functional relationships is relatively small (see Bub and Clifton 1996).

---

[14] Such arguments are typically made in the context of Bohm's theory, by proponents of the theory.

[15] Every physical quantity of a system that can be expressed as a function of the positions and momenta of its particles and its Hamiltonian are determinate properties of the system in classical mechanics. This is what I mean by the rich set of classically determinate properties.

One might insist that all that it really takes to explain our experience is that there be *some* determinate physical quantity in our theory that is well-correlated with our experience. It seems to me, however, that while the supposed association between the determinate physical quantity and our experience *might* explain our determinate experience, there remains the question whether it in fact *does* explain our determinate experience given our actual physiology and practice. It is not that there is a definitive argument that none of the determinate quantities that have been proposed so far (like position in Bohm's theory) can be right; rather, it is that having a physical quantity that is well-correlated with our experience is not the same as having good empirical evidence that it is that particular quantity that would in fact explain the determinateness of our most immediate measurement records.

If this is right, then along with its other problems we are apparently faced with the embarrassment of having to stipulate a special privileged physical quantity as always determinate in the splitting-worlds theory (just as one must in Bohm's theory).

6.3   *Many worlds without splitting*

Many of the problems encountered by the splitting-worlds theory can be avoided by identifying worlds with trajectories of local states rather than with local states at a time. That is, one might stipulate that the global state always evolves in the usual linear way, choose a preferred basis that makes observers' records always determinate in each local state (the states described by the terms in the expansion), then choose a connection rule for hooking up local states at different times into trajectories. Each of these trajectories might then be taken as representing a history of a possible world. Finally, one might specify a probability measure over possible histories (hence over the possible worlds) that determines the prior epistemic probability that each possible world is actually ours (Fig. 6.2).

Call this a *many-threads* theory. There is a different many-threads theory for each choice of preferred basis, connection rule, and initial probability measure. On such a theory, the connection rule determines what threads there are, and one might think of each thread as the specification of the history of a single world. These worlds never split, so there are no special problems with the identity of observers over time. Consequently, it seems to me that one could take each physically possible world described

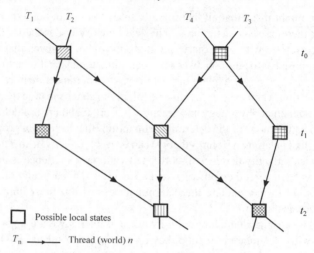

FIG. 6.2 How a connection rule might thread possible local states into worlds.

by such a theory to be real while only one of these worlds would be *ours* (with prior epistemic probabilities given by the measure over trajectories stipulated by the theory). One can see how a many-threads theory works by comparing a repeat *x*-spin measurement on a splitting-worlds theory (Fig. 6.3(*a*)) with the same sort of measurement on a (memory-preserving) many-threads theory (Fig. 6.3(*b*)).

The connection rule ties together states of the same world at different times so that one can make sense of the transtemporal identity worlds and hence observers, which in turn makes it easier to interpret probabilities in the theory. On the other hand, one must still choose a preferred basis (or, more generally, a preferred basis for each time) and a connection rule for threading together local states at different times.[16] So we still have the preferred-basis problem and we need to find a rule for connecting local states so that the theory is both empirically coherent and empirically adequate.

There is an analogy between possible worlds in a many-threads theory and possible worlds in classical mechanics. For a given Hamiltonian, each initial state in classical mechanics picks out a different trajectory through

---

[16] It would be much better if we had a natural rule that selected a single objectively preferred set of alternative histories where observers have determinate records given the boundary conditions of the universe. This is what Gell-Mann and Hartle wanted in their many-histories theory. I shall consider this in Ch. 8.

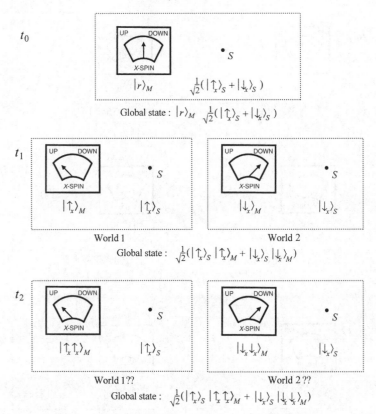

FIG. 6.3(*a*) A repeated *x*-spin measurement in a splitting-worlds theory.

phase space, a different history. One might take there to be a possible world corresponding to each such history and then take these classically possible worlds to be real (in the sense that people sometimes take possible worlds to be real). Finally, one might take a given probability measure over these possible worlds as providing prior epistemic probabilities for each world being ours, where *ours* is used here in its perfectly ordinary sense (since one can now appeal to standard notions of personal identity). Analogously, one might take each history described by a particular many-history formulation of quantum mechanics as the history of some real world (in whatever sense one takes possible worlds to be real), and then take the probability distribution over these worlds as providing prior epistemic probabilities for each world being ours.

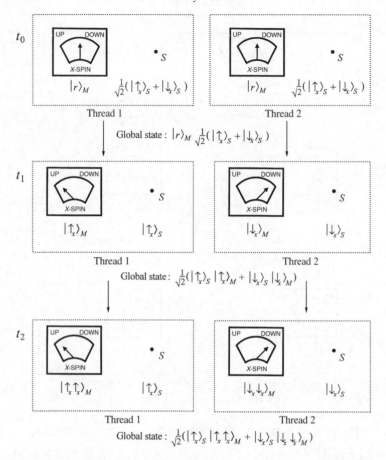

FIG. 6.3(*b*) A repeated *x*-spin measurement in a many-threads theory.

There is, however, also a disanalogy between such classical and quantum worlds. Given a Hamiltonian, the possible classical worlds are indexed by the set of possible initial states—that is, the initial state of a classical world fully determines the history of that world. But the initial local state of a quantum world would typically not determine that world's history since many different histories might pass through precisely the same local state at a particular time. Indeed, the local state of a world by itself would not even allow one to calculate the quantum probabilities for future states of that world (one would also need to know the global state of the universe).[17]

If one insists on having the physical state of a quantum world at a time provide the quantum probabilities for future states of that world (which, as discussed earlier, might be taken as part of what it would mean for a state to be the state of a *world*), then one would want to include the global wave function as part of the state description of each world. Given both the local state of a world and the global wave function, one could use the linear dynamics and the particular connection rule stipulated by one's theory to make predictions about the future states of one's own world. If one did not know the complete local state of one's world, then one might use the linear dynamics, the connection rule, and the prior epistemic probabilities over worlds stipulated by the theory to make statistical predictions.

And this should be starting to sound familiar. If we include the global state as a part of the state description of each world, then one might think of each world as being described by a hidden-variable theory. If the preferred basis is position, if the connection rule is that any local state can follow any other with probabilities given by the squares of the coefficients on the current global state, and if the initial epistemic probabilities are given by the squares of the coefficients on the initial state, then each world is described by Bell's Everett (?) theory.[18] If the preferred basis is position, if the connection rule is Bohm's velocity equation, and if the initial epistemic probabilities are given by the squares of the coefficients on the initial wave function (as stipulated by the distribution postulate), then each world is described by Bohm's theory. More generally, the preferred basis stipulated by a many-threads theory determines what physical quantity is always determinate in a corresponding hidden-variable theory, and the connection rule gives the dynamics for this quantity. That is, if one includes the global wave function in the state description of the worlds, then each world might

[17] That is, as long as one allows for the possibility of interference between the various histories (as happens in the context of A-type measurements, for example), then the current local state of a history is not enough to determine its future local state. Whether or not there will in fact be interference effects between histories depends upon the global state and the Hamiltonian.

[18] One might think of the probabilities in Bell's Everett (?) theory dynamically, as Bell does. On this view, one simply thinks of the next local state as being chosen randomly as the local state evolves: the probability of the next local state being ψ_i is $|a_i|^2$, where ψ_i is a term in the wave function when written in the configuration basis and a_i is the coefficient on this term. It is easy to see that this is empirically equivalent to the other way of picturing the theory.

be thought of as being described by a particular hidden-variable theory, where the preferred basis selects the always determinate physical quantity (the hidden variable), the local state of each world at a time gives the value of this quantity in that world, and the connection rule (together with the linear dynamics) determines, in so far as it is determined, how the quantity evolves in each world: *a many-threads theory is ultimately just a hidden-variable theory where one simultaneously considers all physically possible worlds.*

Bell's Everett (?) theory illustrates the role that the connection rule plays in a many-threads theory. If any local state can follow any other local state, then one might be able to argue that most observers' 'records' would be consistent with the statistical predictions of quantum mechanics, but one could have no empirical grounds for choosing one set of dynamical laws over another. The connection rule must be such that it makes our memories reliable in order for us to have any empirical justification for accepting the linear dynamics and that particular connection rule (the auxiliary dynamics). One can get an empirically coherent theory by stipulating a connection rule that typically makes records of past events reliable in a world. Bohm's theory does precisely this and in a perfectly natural way.

The moral then is this: if one tries to fix the splitting-worlds theory by getting rid of the splitting process (which is clearly the source of the identity and probability problems faced by the splitting-worlds theory), and if one is willing to think of the global wave function as a part of the state of each world, then one naturally ends up with a perfectly ordinary hidden-variable theory—and Bell's Everett (?) theory and Bohm's theory ultimately do not look like such bad ways of understanding Everett after all. Such theories are easily understood, and we know that we can get them to make essentially the same empirical predictions as the standard theory does for the always determinate quantities (as we shall see in the next chapter). Perhaps interpreters of Everett would make their own positions clearer by contrasting them with a hidden-variable theory that is taken as simultaneously describing the time-evolution of many possible worlds.

7

MANY MINDS

ONE might understand Everett's branches as describing the states of different *minds* rather than different *worlds*.[1] On this sort of theory an observer's determinate experiences and beliefs are explained by the fact that he always has a determinate mental state. In this sense, such theories might be said to predict appearances directly. Albert and Loewer have done the most to explain how such a theory might work. I shall also discuss Lockwood's many-minds theory and its relationship to Albert and Loewer's. And finally, I shall discuss the relative-fact theory and the correlations-without-correlata view.

7.1 *How many minds?*

Albert and Loewer (AL) have sketched two closely related theories (AL 1988). I shall call one the single-mind theory and the other the many-minds theory. These theories provide a way of understanding what Everett meant when he claimed that his goal was to recapture the usual predictions of quantum mechanics *as subjective appearances*.

Like Bell's Everett (?) theory and Bohm's theory, both the single-mind and many-minds theories can be thought of as attempts to fix the bare theory. The main problem with the bare theory again is that while it predicts that one will typically believe that one got a determinate result after a successful measurement, there is typically no determinate result that one believes that one got. Perhaps the most straightforward way to ensure that observers always have determinate experiences and beliefs (and in some sense the most *ad hoc*) is simply to stipulate that observers do in fact always have determinate mental states. This is precisely what Albert and Loewer do. While they take the usual linear dynamics to provide a complete and accurate description of the time-evolution of the *physical* world, they postulate the existence of *nonphysical* minds that always have determinate mental states whose time-evolution is given by a mental dynamics.

---

[1] As suggested by Richard Healey (1984: 608, 612–14) and others.

7.1.1   *The single-mind theory*

If one takes the linear dynamics to be universally true, as Everett insisted, then the physical state of an observer and his object system after a measurement will typically be an entangled superposition of the observer having recorded mutually contradictory results, and there will be nothing in the quantum-mechanical state that picks out a single, determinate measurement result. Consequently, if one wants a complete description of the state of affairs to pick out a determinate result, then one must add something to the usual quantum-mechanical state. If one understands the measurement problem as a problem in accounting for the determinate experiences and beliefs of observers, then the most direct way to solve the problem would be to stipulate that observers always have determinate mental states. A complete specification of the state, then, would require one to supplement the usual quantum-mechanical state with a description of the mental states of all observers.

The single-mind theory is a sort of hidden-variable theory, but instead of taking positions as always determinate, one takes mental states as always determinate. While in Bohm's theory one might worry that making positions determinate may not ultimately explain our determinate experiences and beliefs, there is no such worry here. By making mental states always determinate, one automatically accounts for the determinateness of our experiences and beliefs (assuming that mental states determine both experiences and beliefs), and if these mental states evolve in the right way, then the theory will be empirically adequate. While one might complain that this solution of the determinate-experience problem is cheap, it is none the less effective.

Let M be an observer, and let $B_M[x](t)$ represent that M believes x at time t. The single-mind theory requires only a partial correspondence between physical and mental states. It is assumed that there is generally a physical state $\psi_M[x](t)$ that implies that M believes x at time t, $B_M[x](t)$. But it is also an important feature of the theory that $B_M[x](t)$ does not necessarily imply that the quantum-mechanical state is $\psi_M[x](t)$. It is this second feature that allows an observer's *physical state* to be a complicated superposition of physical states corresponding to mutually contradictory beliefs concerning the outcome of a specific measurement while his *mental state* is always one where he has a determinate belief corresponding to exactly one element of the superposition. Since $\psi_M[x](t)$ entails $B_M[x](t)$, one might think of this physical state as an eigenstate of belief; in this case, an eigenstate of M believing x. Similarly, one might suppose that it is always possible to represent the universal state as a superposition of states that describe M as having different but determinate beliefs, and that M's mental state is always associated with exactly one of the terms in such an expansion.[2]

The physical state always evolves in the usual linear way, but in order to have a complete theory, we also need to specify a mental dynamics. Perhaps the simplest mental dynamics would be one like the configuration dynamics in Bell's Everett (?) theory, where the probability of M's current mental state being $B_M[x](t)$ is fully determined by the current physical state alone and is always equal to the norm-squared of the component of the universal state that describes M as currently believing x. One might picture the mental state of an observer as randomly jumping from branch to branch of the determinate-belief superposition, paying no heed to where it was last, in such a way that the probability of it being associated with a particular branch is always equal to the norm-squared of the amplitude of that branch. This is the single-mind analogue of the evolution of position in Bell's Everett (?) theory.

If M measured the x-spin of a system S initially in an eigenstate of z-spin, then, by the linear dynamics and by what it means to be a good observer, M would end up in a superposition of belief eigenstates corresponding to getting mutually incompatible results:

$$\psi_{M+S}(t_1) = \frac{1}{\sqrt{2}} \left[\psi_M[\uparrow]|\uparrow\rangle_S + \psi_M[\downarrow]|\downarrow\rangle_S \right]. \quad (7.1)$$

Here one component of his physical state describes him as believing that the result was x-spin up and the other describes him as believing that the result was x-spin down. On the single-mind theory, however, M's mental state would end up associated with exactly one term in the post-measurement superposition—in this case, there would be an even chance that M's post-measurement mental state would be $B_M[\uparrow](t_1)$ and

---

[2] This explanation of the determinacy of experience requires one to adopt a theory of mind where an observer's mental state is well-defined at an instant. Matthew Donald (1996) has argued that such an assumption is implausible. It seems to me, however, that if one was willing to go along with Everett's dispositional account of mental states initially, then one should be willing to go along with the assumption that a mental state is well-defined at an instant (since dispositions are well-defined at an instant). If one wants an account of mental states in terms of mental processes, then more work needs to be done, but the problem does not seem to me to be insurmountable. First, instead of the determinate-belief basis, one might try choosing the preferred basis to be whatever basis makes a mental process determinate, then show that the mental dynamics generates sequences of process states that can be identified with the observer having a determinate mental state.

that it would be $B_M[\downarrow](t_1)$. If M repeats his measurement, then ideally he would end up in the physical state

$$\psi_{M+S}(t_2) = \frac{1}{\sqrt{2}} \left[\psi_M[\uparrow, \uparrow] | \uparrow \rangle_S + \psi_M[\downarrow, \downarrow] | \downarrow \rangle_S \right]. \qquad (7.2)$$

And on the simple mental dynamics described above, there would again be an even chance of getting each result. This means that there is an even chance that M would end up in a mental state where he believes that he got x-spin *down* as the result of his second measurement even when he in fact got x-spin *up* as the result of his first measurement. But if M does get a different result for his second measurement, he will falsely 'remember' getting precisely that result for his first measurement: if M ends up believing that he got x-spin down for his second measurement, his mind will also be associated with a branch of the wave function where he believes that he got x-spin down for his first measurement *regardless of what he actually got*. So even if the mental dynamics is a random function of the observer's current physical state alone and if the observer's memories were consequently unreliable concerning even his own past mental states, the observer would never notice (Fig. 7.1). The linear dynamics, the properties of a good observer, and the relationship between mental states and physical states would conspire to make it impossible for an observer to tell that his experiences were in fact pathologically discontinuous. Indeed, on this dynamics, just as in Bell's Everett (?) theory, an observer would almost always judge that his measurement results exhibited the usual quantum statistics.

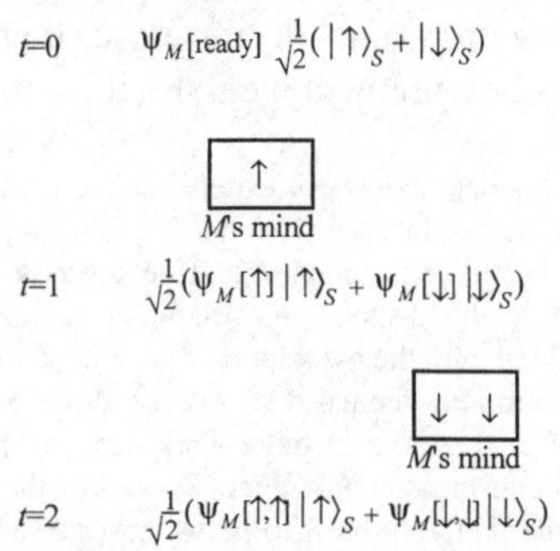

FIG. 7.1 A randomly jumping mind.

Also, just as in Bell's Everett (?) theory, the fact that an observer's records of past events would be unreliable makes it difficult to say how the observer could have any empirical justification for accepting any dynamical law, even the single-mind theory's dynamics, if this version of the single-mind theory were true. In order to avoid the problem of empirical coherence, one might want to specify a mental dynamics where the current mental state is a function of both the current quantum-mechanical state *and past mental states*. While they do not provide a complete mental dynamics, Albert and Loewer want their dynamics to have this property of reliability.[3] What they in fact use is a mental dynamics that mimics the empirical predictions of the standard collapse theory in so far as this is possible. The intuition is that a mind sticks with a branch in the determinate-belief decomposition of the global state until a measurement occurs in that branch. The mind is then assigned to a new branch just as if the physical state really was described by the branch that it was on and a collapse occurred with the probabilities given by Born's rule. The observer's new mental state will typically be one where his memories are reliable. While this is not a complete mental dynamics, they are sure that one can cook up a dynamics that will make the single-mind theory both empirically coherent and empirically adequate (Albert 1992: 129).

AL do not like their single-mind theory because it does not generally predict that mental states would supervene on physical states since even a complete description of the physical world would fail to determine the mental state of an observer. The physical state $\psi_{M+S}(t_1)$, for example, is consistent with either $B_M[\uparrow](t_1)$ or $B_M[\downarrow](t_1)$. AL describe this type of nonphysicalism as 'especially pernicious', and they tell us that it is this lack of mental supervenience that leads them to consider the many-minds theory (AL 1988: 206). As Albert put it:

the dualism of this sort of picture is ... pretty bad. On this proposal (for example) all but one of the terms in a superposition [like those above] represent (as it were) *mindless hulks*; and *which one* of those terms is *not* a mindless hulk can't be deduced from the physical state of the world, or from the outcome of any sort of experiment; and it will follow from this proposal that most of the people we take ourselves to have met in our lives have as a matter of fact *been* such hulks, and not really people (not really animate, that is) at all! (1992: 130)

The problem, then, is this: when I make observations and then compare notes with a friend, his mind and my mind will each be associated with

<hr>

[3] See Albert (1992: 126–9) and Loewer (1996). I shall discuss this second paper later in this chapter.

some term in the determinate-belief expansion of the global state, but not necessarily the same term. Suppose that my friend (F) and I (M) both make perfect x-spin measurements and end up in the physical state

$$\psi_{M+S}(t_2) = \frac{1}{\sqrt{2}} \left[\psi_M[\uparrow]\psi_F[\uparrow]|\uparrow\rangle_S + \psi_M[\downarrow]\psi_F[\downarrow]|\downarrow\rangle_S \right]. \quad (7.3)$$

Suppose, however, and this is entirely possible given the mental dynamics that AL describe, that my mental state ends up associated with the first term in the superposition and I get x-spin up, and that my friend's mental state ends up associated with the second term and he gets x-spin down. Since the mental dynamics is supposed to be memory-preserving, in so far as is possible, when I ask my friend what he got as the result of his x-spin measurement, my mental state will end up associated with a branch where I still believe that I got x-spin up, but because of the correlation between my physical state and my friend's, this will also be a mental state where I believe that he told me that he also got x-spin up. I might conclude, from what I heard, that my friend believes that he got x-spin up, but this would be false. While I am getting what I take to be perfectly clear answers to my questions, these answers have nothing at all to do with the actual mental state of my friend, and I am consequently not talking with my friend at all. To the extent that I am talking with anyone, I am talking with a mindless hulk. And my friend will find himself in a similar situation (Fig 7.2). Indeed, while observers might believe that they talk with each other routinely, since the physical state does not do much to constrain the mental state of an observer and since observers fail to have direct epistemic access to each other's mental states, it would

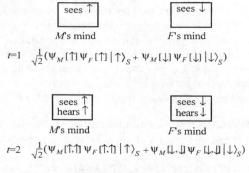

FIG. 7.2 Talking to mindless hulks.

typically be impossible to figure out what anyone else actually believes. So while the single-mind theory can account for each observer's experiences in an empirically adequate way, it makes real communication between observers impossible and thus describes a sad and lonely world, unless one takes comfort in the company of mindless hulks.

One solution to the mindless-hulk problem would be to change the mental dynamics so that all minds end up associated with the same branch of the wave function (written in the determinate-belief basis). The idea here is that since the time-evolution of my mental state already depends on my past mental states to provide for reliable memories, one might as well have it also depend on other observers' mental states to provide for genuine communication between observers. When I get x-spin up, then all other observers become associated with the branch that determines the outcome of my observation (Fig. 7.3).

This solves the mindless-hulk problem, but on this modified single-mind theory the result of my measurements might instantaneously determine the outcome of measurements made by a distant observer (in an EPR experiment, for example, where one thinks of each observer's mind as having the same location as his body). Note that this is not the case on AL's mental dynamics since on their dynamics each mind evolves *independently* of all other minds. In an EPR experiment, on AL's mental dynamics, each observer would end up believing that his results and his friend's results were statistically correlated in such a way as to violate the Bell inequalities, but they would as a matter of fact not know what the other observer's results were. What each thought the other said his results were would be determined by the term that their own mind ended up associated with, and since we know from the bare theory that most terms,

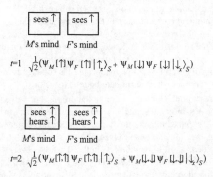

FIG. 7.3 A mental dynamics without mindless hulks.

in the norm-squared measure, will exhibit the usual EPR correlations, and since this measure represents the probability of an observer's mind ending up associated with each term, with high probability an observer's mind will eventually be associated with a term that exhibits the usual EPR correlations.[4]

While one might not like the nonlocality (more specifically, the lack of Lorentz-covariance) of the modified mental dynamics where the evolution of each mind depends on the current state of other minds (just as the evolution of a particle's position depends on the positions of other particles in Bohm's theory), when compared with other options for solving the mindless-hulk problem, it may not seem so bad.[5]

7.1.2 *The many-minds theory*

AL's solution to the mindless-hulk problem is to suppose that every sentient physical system, every observer, is associated with not a single mind but rather a continuous infinity of minds (AL 1988: 206). Each mind is supposed to evolve exactly as described in the single-mind theory; there are just more of them associated with each observer (Albert 1992: 130). Each of an observer's minds evolves independently of his other minds, but in such a way that a mind's own beliefs about its own past mental states are typically reliable. Whenever a measurement is made, some of an observer's minds will become associated with each branch of the wave

---

[4] There is a close relationship between the AL single-mind theory and the bare theory. Since the probability of a mind being associated with a particular term on AL's mental dynamics is equal to the norm-squared of the coefficient on that term and since the bare theory's general limiting property tells us that as the number of experiments gets large almost all of the terms, in the norm-squared measure, in the determinate-belief expansion of the state describe situations where an observer's results exhibit the usual quantum statistics (including the joint and conditional statistics and the statistical correlations with other observers' results), an observer on the single-mind theory will *almost always* believe that his own results and the results of his colleagues exhibit the usual quantum statistics. Even though, because he has most likely been talking to mindless hulks, he will not actually know what the results of his colleagues were (because his colleagues' minds are most likely associated with other terms in the determinate-belief expansion).

[5] Lockwood raises a nonlocality objection to the proposal of solving the mindless-hulk problem by fiat: 'The possibility of actually *meeting* mindless hulks can perhaps be ruled out by *fiat*. For one could stipulate that the stochastic evolution takes place under the constraint that living bodies, as they manifest themselves within a single Everett branch, are required either all to be ensouled, or all to be mindless hulks. That, however, would call for a *nonlocal* coordination of the various mind evolutions, and thus undercut what is one of the major motivations for adopting a no collapse, no hidden variables view, in the first place' (1996: 175).

function that describes the observer as getting a definite result. Since the norm-squared of the coefficient on each term in the determinate-belief expansion of the wave function determines the probabilities for a particular mind being associated with that term, one might naturally take the norm-squared of the coefficient on each term to determine the *measure μ* of the continuous infinity of an observer's minds associated with that term. And if one believes that another observer reported that he also got *x*-spin up, for example, there will typically be at least some of that observer's minds which did in fact get *x*-spin up, and it is these minds that one is talking to. It is as if every hulk has been given a mind; indeed, many.

One might think of each of an observer's minds as representing a different perspective or view of the physical world. But while each mind sees a single, determinate, and consistent series of events, the global observer, as a collection of minds, has many mutually incompatible experiences. If an observer begins in an eigenstate of being ready to make an *x*-spin measurement of a system in an eigenstate of *z*-spin, all of the observer's minds will end up with determinate beliefs concerning the result of his observation, but not the same determinate beliefs. Here, with probability one, half of the observer's continuous infinity of minds would end up believing that the result was *x*-spin up and the other half would end up believing that the result was *x*-spin down.

AL describe their many-minds theory as follows:

The individual minds, as on the [single-mind theory], are not quantum mechanical systems; they are never in superpositions. This is what is meant by saying that they are non-physical. The time evolution of each of the minds on the [many-minds theory] is, just as on the [single mind theory], probabilistic. However, unlike the [single-mind theory], there are enough minds associated with the brain initially so that minds will end up associated with each of the elements of the final superposition. An infinity of minds is required since a measurement or a sequence of measurements may have an infinite number of outcomes. Furthermore, although the evolution of individual minds is probabilistic, the evolution of the set of minds associated with [a particular observer] is deterministic since the evolution of the measurement process is deterministic and we can read off from the final state the proportions of the minds in various mental states. (AL 1988: 207)

It is important to be clear here about exactly what is evolving in a deterministic way and what sort of supervenience the many-minds theory provides. One might distinguish between at least three types of mental states in the theory. Call the state of one of an observer's minds a *local mental state*. Local mental states evolve randomly (but in a memory-preserving way) according to the dynamics described by the single-mind theory and do not supervene on observers' physical states. Call the state of all of an observer's minds his *complete mental state*. This state tells us *which minds* are in *which local states* (we are assuming with AL that minds have transtemporal identities). It also evolves randomly as the local mental states of each mind evolves. These states do not supervene on observers' physical states either. Finally, call the measure of an observer's minds associated with each term in the determinate-belief-basis expansion of the wave function the observer's *global mental state*. It is only the observer's global mental state that one might expect to evolve in a continuous deterministic way and to supervene on his physical state (though it might happen, since the dynamics of each mind is stochastic, that all of an observer's minds jump to a single branch. But events like this would be expected with probability zero).

Despite its inherent craziness, AL argue that the many-minds theory has several advantages over other interpretations of Everett and other versions of quantum mechanics generally. First, unlike the bare theory, the many-minds theory is 'in accord with our very deep conviction that mental states never superpose' (AL 1988: 208). None the less, AL argue that it remains true to Everett's fundamental idea that the time-evolution of the entire universe and every *physical system* is completely and accurately given by the linear dynamics: 'There is no need to postulate collapses or splits or any other non-quantum mechanical *physical* phenomena' (208). The reason that *physical* is emphasized here, of course, is that the many-minds theory supposes the existence of *nonphysical* entities, the observer's minds, whose local states are not determined by the global quantum-mechanical state, and it is these nonphysical entities that exhibit what one might call nonquantum mechanical behaviour. On the other hand, it seems that Everett really did believe that mental states evolve in a fundamentally different way than physical states. He repeatedly said that, on his formulation of quantum mechanics, while the evolution of the universal quantum-mechanical state is always linear and deterministic, the evolution of subjective appearances is random and probabilistic. The many-minds theory also provides a particularly natural way to make sense of Everett's claim that, at the end of a typical measurement, while in one sense there is only one physical observer, in another sense there are many observers with mutually incompatible experiences: if we count physical systems in the many-minds theory, there is only one observer; but if we count minds, then there are many with mutually incompatible experiences.

Another virtue of the many-minds theory is that 'the choice of basis vectors in terms of which the state of the world is expressed has no physical significance' (AL 1988: 208). AL take this to mean that the many-minds theory avoids the preferred-basis problem. This is not to say that the many-minds theory has no preferred basis—after all, the determinate-belief basis acts as a sort of preferred basis in the theory. While one might say that this basis has no *physical* significance and that it consequently plays a more modest role than the preferred basis in the splitting-worlds interpretation (it does not, for example, determine once and for all what physical properties are determinate), the determinate-belief basis plays an important empirical and explanatory role in the theory. One must, for example, appeal to the determinate-belief basis in order to specify the transition probabilities for a mind to go from one mental state to another. More generally, without explicitly specifying the preferred basis, the many-minds theory has no explicit mental dynamics. And since the preferred basis is nowhere explicitly specified, one might argue that the many-minds theory is at best incomplete.[6]

AL also argue that the many-minds theory encounters no special problems in interpreting probability: 'Probabilities are completely objective, although they do not refer to physical events but always to sequences of states of individual minds' (AL 1988: 208). Since each mind follows a random trajectory with probabilities given by the mental dynamics, one should eventually expect the memories of almost all of an observer's minds, in the norm-squared measure, to exhibit the usual quantum statistics. Further, on the sort of mental dynamics that AL have in mind, probabilities are, as in the standard collapse formulation of quantum mechanics, the result of fundamental stochastic events.

AL also claim that the many-minds theory's dynamical laws can be expressed in a local, Lorentz-covariant form, which means that the many-minds theory meshes well with relativity (AL 1988: 209–10). What they have in mind here is that since one can read off the global mental state of an observer from the universal wave function, and since the evolution of the wave function can be expressed in a covariant form, so can the global mental dynamics.

---

[6] Like Albert and Loewer, Lockwood's many-minds theory also requires a preferred basis where every observer has determinate beliefs. Each Everett branch then can be described by a product of a determinate-belief state and the corresponding relative state of the composite system containing the observers. Also like Albert and Loewer, Lockwood argues that there is no '*objectively* preferred basis': 'For a many minds theorist, the *appearance* of there being a preferred basis, like the *appearance* of state vector reduction, is to be regarded as an illusion. And both illusions can be explained by appealing to a theory about the way in which *conscious mentality* relates to the physical world as unitary quantum mechanics describes it' (1996: 170).

Finally, AL consider the many-minds theory to have a decided advantage over the single-mind theory because the many-minds theory allows an observer's global mental state to be uniquely fixed by his physical state. As they describe the deal, 'We have purchased supervenience of the mental on the physical at the cost of postulating an infinity of minds associated with each sentient being' (AL 1988: 207). In the end, however, one might take the price to be too high considering what one gets.[7]

Consider the type of mental supervenience that the many-minds theory provides. If one wants mental supervenience, one presumably wants the mental state that one is capable of introspecting right now, the mental state that one has epistemic access to, to supervene on one's physical state. I believe that I have a more or less definite mental state characterized by a single set of more or less consistent beliefs. But the many-minds theory tells me that I (global) am associated with an infinite set of minds that most likely have wildly contradictory beliefs and whose mental states I (local) generally cannot know. What comfort is it supposed to give me that my global mental state supervenes on my physical state when I don't even know what my global mental state is? If one wants supervenience, then one presumably wants one's *local* mental state, the state that determines one's experiences and beliefs and the only state to which one has epistemic access, to supervene on one's physical state. And in the many-minds theory my local mental state does not supervene on my physical state.

That each observer is associated with a continuous infinity of minds is at least counter-intuitive. Being counter-intuitive is not a fatal flaw in a theory, but one might eventually decide that associating each observer with an infinity of minds costs more than it's worth. If someone is worried about the mindless-hulk problem, then supposing, without any possibility

---

[7] Again, an observer's global mental state does not *necessarily* supervene on his physical state. The mental state of each of the observer's minds is a random function of his physical state and independent of the states of his other minds, and his global mental state is determined by the local mental states of all of his minds. Consequently, it could happen, although it would be extraordinarily unlikely, that all of the observer's minds, for example, become associated with a single branch in the determinate-belief expansion of the wave function. I do not take this lack of strict supervenience as a serious problem, but if one is worried about it, one could guarantee strict supervenience by changing the theory. Rather than have each mind evolve randomly, associate one mind with every possible memory-preserving trajectory through an observer's determinate mental states, let μ^* be the product measure on the set of minds generated by the Albert and Loewer measure μ on minds at a time, and take μ^* as determining the probability that a local observer will find himself associated with one of the minds (μ^* would play the same role in this theory as the measure over trajectories with reliable records in the many-worlds theories discussed earlier). Here the wave function would strictly determine the global mental state of the observer and how it evolves.

of empirical support, that every hulk has at least one mind associated with it does not seem to me to be the sort of thing that should resolve the worry. Rather, the many-minds theory has an *ad hoc* flavour to it. It looks as if minds were added just to give hulks mental states—and they were.

The difference between a many-worlds theory (like the many-threads theories at the end of the last chapter) and a many-minds theory (like AL's) is not that great. It really just amounts to a difference in *what sort of facts* one takes as being determinate. On a many-worlds theory physical records and thus local mental states are determinate, and on a many-minds theory it is only local mental states that are determinate. If one had a successful many-minds theory, then one could always convert it to a many-worlds theory. By doing so, one could solve the mindless-hulk problem and eliminate the mind–body dualism to which AL are committed at the cost of introducing the determinate-belief property as a preferred *physical property*. A many-minds theory with the completely random mental dynamics that were discussed first would convert to something like Bell's Everett (?) theory but where one considers all possible trajectories and, rather than position always being determinate, it would be *belief* that is always determinate (*belief* here is supposed to be the physical property that if determinate would make all observers' beliefs determinate). AL's many-minds theory would convert to a many-worlds theory where the physical property that determines one's beliefs is always determinate and each world evolves independently of the others in a random, but memory-preserving, way similar to the evolution of beliefs predicted by the standard collapse theory.

7.2   *The auxiliary dynamics*

AL did not give an explicit mental dynamics, but they did not take this to be a very serious problem. As Albert put it: 'What's been said so far ... doesn't amount to a completely general set of laws of the evolution of mental states; but laws like that can be cooked up, and they can be cooked up in such a way as to guarantee that everything I've said about them so far will be true' (1992: 129). But just as the many-worlds theories need an explicit rule to connect states at different times, without an explicit mental dynamics, the single-mind and many-minds theories are incomplete. Regardless of whether one likes the ontology of worlds or minds better, one needs an auxiliary dynamics that describes how the determinate local states evolve while the universal wave function is evolving in its usual linear way. Of course, there is always Bell's Everett (?)

solution, but if such a dynamics actually described how worlds or minds evolved, then it would be impossible to tell a convincing story how one ended up with empirical justification for believing that it did. Here I want to discuss two very different ways in which one might go about specifying a dynamics that provides for reliable memories of past events.

7.2.1   *The transcendental approach*

There is a way to exploit the bare theory's suggestive properties in order to get a dynamics for mental states. It is inherently crazy, and it encounters serious problems, but it is perhaps as close as one can get to Everett's original goal of *deducing* the usual predictions of quantum mechanics as subjective appearances from pure wave mechanics.

Consider the single-mind theory. An observer's mind does nothing to affect his physical dispositions. Just as in the bare theory, the physical dispositions of all physical systems, including observers, are fully determined by the wave function and the linear dynamics. And this means that an observer in the single-mind theory will have exactly the same physical dispositions as an observer in the bare theory, which means that all of the conclusions that we reached in the bare theory about what reports an observer would make apply equally well to an observer in the single-mind theory. So assuming that an observer will report and believe whatever he is in an eigenstate of reporting and believing, the bare theory's suggestive properties tell us something about what an observer would report and believe about his own experience and its relationship to the experience of other observers in a single-mind theory. So if we suppose that an observer's reports and beliefs about his own experiences and beliefs are typically true, then the bare theory's suggestive properties strongly constrain the fundamental nature of mental states and how they could evolve in a single-mind theory.

The type of scientific knowledge that we usually take ourselves to have requires our beliefs concerning own experience and the experience of other observers to be typically reliable; indeed, one might take the general reliability of our beliefs about the nature of our own experience to be a precondition for the possibility of scientific inquiry. In any case, let's suppose that whenever a careful inquirer has a sure-fire disposition to report something about his own (or other's) experiences and beliefs, then that report is typically true.[8]

---

[8] This is the transcendental part: one starts by supposing that scientific inquiry is possible—that our beliefs about our own and others' measurement results are typically true.

Suppose an observer makes a perfect x-spin measurement of an object system initially in an eigenstate of z-spin and ends up in the physical state predicted by the linear dynamics. The bare theory's determinate-result property tells us that the observer will report that he got either x-spin up or x-spin down as the result of his measurement. If we stipulate that the mental dynamics is such that an observer's reports concerning his own experiences and beliefs are typically true, then *this means that he must actually end up with a mental state corresponding to one or the other of the two possible x-spin results*. This might be taken as something of a justification for introducing determinate ordinary mental states—the linear dynamics together with Everett's model of observers as automata tell us that an observer will report that he has a determinate ordinary mental state, and by giving the observer a determinate ordinary mental state in the single-mind theory, we make his report true.

Now suppose the observer got the result x-spin up to his first measurement, and then, without disturbing the object system, he makes a second perfect x-spin measurement. The bare theory's repeatability property tells us that an ideal observer will report that he got the same result for both measurements. In order for his belief concerning his own experience to be true here, *the mental dynamics must be such that the observer in fact gets the same result for his second measurement as he did for his first*; that is, he must get x-spin up for his second result. In other words, if one assumes an ideal observer's reports about his experience to be true whenever possible, then the repeatability property requires the mental dynamics to be memory-preserving.

Now suppose that two observers make perfect x-spin measurements. The first observer measures the x-spin of an object system in an eigenstate of z-spin and gets the result x-spin up; the second observer then measures the x-spin of the same object system, which has not been disturbed since the first measurement, and gets a result. The bare theory's agreement property tells us that if the two observers compare their x-spin results, they will conclude that they got the same result; so by assuming that an observer's beliefs about the relationship between his own experience and

The bare theory's suggestive properties then strongly constrain the mental dynamics of the single-mind theory. That is, if an observer's reports about his own experience and its relationship to the experience of other observers are typically true whenever he is in an eigenstate of making such a report, then the suggestive properties tell us what properties the observer's mental state must have and how it must evolve, inasmuch as they tell us what the observer will report about his own experience and its relationship to the experience of other observers.

the experience of other observers are typically true, *the mental dynamics must be such that the second observer in fact gets the same result as the first*, x-spin up. Since the result of the first observer determines the result of the second observer, neither observer has to worry that he is talking to a mindless hulk when he hears his friend tell him that their results agree.

Finally, suppose that an observer performs an experiment where he measures the same observable on a series of systems in identical states. The bare theory's general limiting property tells us that the composite state of the observer and the object systems will approach an state where the observer has the sure-fire disposition to report that his results were randomly distributed with the usual quantum relative frequencies. Note that the general limiting property places an extremely strong constraint on the mental dynamics here: *it requires the mental dynamics to yield the usual quantum statistics for the subjective experience of every observer in the limit*. In order for this report to be true, the mental dynamics must be such that the observer's measurement results really were randomly distributed with the appropriate relative frequencies. Specifically, if one requires that the mental dynamics is trial-independent (requires that it be given by a trial-invariant law), then in order for the observer to have a probability-one chance of his mental state agreeing with the report that he must make in the limit, each measurement result must be randomly determined by probabilities equal to the limiting relative frequencies.

The upshot is that if we accept Everett's model of observers *and* if we assume that one believes whatever one has a sure-fire disposition of reporting *and* if we assume that one's beliefs about one's own experience and its relationship to the experience of other observers are typically true *and* if we require a trial-independent dynamics, then we can deduce almost all of the predictions of the standard collapse formulation of quantum mechanics as subjective appearances from pure wave mechanics. It is remarkable how far we can push this. Or, as Everett might have put it, it is at least a matter of intellectual interest that pure wave mechanics places this strong a constraint on the auxiliary dynamics. It may even seem, at first glance at least, that we have fully determined the auxiliary dynamics. Here the dynamics that governs the evolution of mental states and thus accomplished Everett's goal of deducing all the empirical predictions of quantum mechanics as subjective experiences of observers.

It is, however, somewhat too optimistic to conclude that this transcendental approach fully determines the auxiliary dynamics. Perhaps the most serious problem is that this approach is based on the assumption that the mental dynamics is such that an observer's beliefs concerning his own experience and its relationship to the experience of other observers are typically true. As the *typically* here suggests, this assumption is ultimately rather vague. And as long as this basic assumption is vague, the transcendental approach cannot determine a complete and unambiguous mental dynamics. So why don't we make it precise and assume that the mental dynamics is such that every observer's reports and beliefs are *always* true?

One problem with this is that if one accepts the bare theory's limiting properties, which play a basic role in the transcendental approach, then this condition is impossible to satisfy (for reasons that I have already discussed, but in another context). Suppose an observer measures the same observable on each of a series of identically prepared systems. The observer will always report that he got a determinate result for each of his measurements, and as the number of measurements increases, the composite state will approach one where he will report that he got the usual quantum statistics. But since the norm-squared of the coefficient on the term in the determinate-belief expansion of the observer's state that describes that observer as having recorded a particular sequence of results always goes to zero in the limit, the composite system will also approach a state where the observer will have the sure-fire disposition to report that he failed to get any particular sequence of results that we ask him about (this is the same feature of the linear dynamics that we considered earlier in the context of the Bohm's theory). This means that if the observer believes what he reports, then in the limit he will believe that he got *some* determinate sequence of results but that he failed to get *each* particular sequence of results, which presumably cannot be true. The moral here is that if one believes that the limiting properties tell us what an observer would believe in the limit, then it is impossible for any no-collapse theory to make all of an observer's reports and beliefs true of even his own experience.

Here is one last observation about the auxiliary dynamics. Suppose a many-minded observer M measures the x-spin of a system S in an eigenstate of z-spin and ends up in the state

$$\frac{1}{\sqrt{2}} \left[\psi_M[\uparrow] |\uparrow\rangle_S + \psi_M[\downarrow] |\downarrow\rangle_S \right]. \tag{7.4}$$

One would expect half of the observer's minds to end up believing that S is x-spin up and half to end up believing that S is x-spin down. Now suppose that someone measures an observable A that has (7.4) as an eigenstate with eigenvalue $+1$ and everything orthogonal as an eigenstate with eigenvalue -1. Measuring A does not disturb the observer's physical state, but on AL's dynamics it does change the states of the minds. Albert stipulates that each individual mind 'always evolves just as if that mind's present beliefs about the quantum state of the world were actually true', and in this situation Albert interprets this rule to entail that half of those minds that believed that S was x-spin up before the A-measurement will believe that it is x-spin down after and half of those minds that believed that S was x-spin down before the A-measurement will believe that S is x-spin up after (1992: 187). In other words, the memories of a particular mind concerning what it (he or she) believed about the x-spin of S before the A-measurement may be false.

But it does not seem to be a necessary property of an empirically adequate theory that an observer's memories be unreliable in a Wigner's-friend experiment. Consider Bohm's theory. Suppose that our friend records his measurement result in terms of the position of something, and suppose that our A-measuring device is designed so that it does not disturb our friend's quantum-mechanical state and so that the effective wave function of its own pointer is stationary if and only if the system being measured beings in a $+1$ eigenstate of A. In this situation an A-measurement of our friend and his object system would produce no probability currents, which in Bohm's theory means that nothing (and this includes the system that our friend used to record his measurement result) would move, which means that our A-measurement would not disturb our friend's record.

7.2.2   *The Bohm–Bell–Vink dynamics*

On Bohm's theory the positions of particles are always determinate. As the wave function evolves in the usual linear way, the theory tells us how positions change. If the particles are initially distributed in the usual random quantum way, then they will always be distributed in the usual way, and the theory makes the usual statistical predictions for position measurements. But if a solution to the measurement problem requires one to account for the determinate experiences of observers, then one will take Bohm's theory to solve the measurement problem only in so far as one is convinced that making positions determinate makes the experiences of sentient beings determinate. And there is no reason to suppose that all sentient beings must ultimately record their measurement results in terms of position. On the other hand, the trick that Bohm used to give a dynamics for position can be used to give a dynamics for almost any property (including whatever property would make beliefs determinate). [9]

Following Bell (1987: 176–7), Vink (1993) has described a natural way to extend Bohm's theory to physical quantities other than position. On Vink's formulation of quantum mechanics the quantum-mechanical state ψ evolves in the usual linear, deterministic way and the Bell–Vink dynamics (which is simply an extension of Bohm's dynamics to discrete quantities generally) describes the time-evolution of the determinate physical quantities. Suppose that the current value of some physical quantity is o_m. The probability that the value jumps to o_n in the time-interval dt is $T_{mn}dt$, where T_{mn} is an element in a transition matrix that is completely determined by the evolution of the wave function. More specifically, the wave function evolves according to the time-dependent Schrödinger equation

$$i\hbar\partial_t|\psi(t)\rangle = H|\psi(t)\rangle, \tag{7.5}$$

where $\hat{H}$ is the global Hamiltonian. The probability density P_n is defined by

$$P_n(t) = |\langle o_n|\psi(t)\rangle|^2 \tag{7.6}$$

and the source matrix J_{mn} is defined by

$$J_{mn} = 2 \operatorname{Im}(\langle\psi(t)|o_n\rangle\langle o_n|H|o_m\rangle\langle o_m|\psi(t)\rangle). \tag{7.7}$$

Finally, if $J_{mn} \geq 0$, then for $n \neq m$

$$T_{mn} = J_{mn}/\hbar P_m; \tag{7.8}$$

and if $J_{mn} < 0$, then $T_{mn} = 0$.

Vink shows that if one takes Q to be position and takes the lattice to be fine-grained enough, then this stochastic dynamics approximates Bohm's position dynamics over an appropriate time interval. More generally, he argues that his theory makes the same statistical predictions for the values of the determinate quantities as the standard theory (whenever it makes coherent predictions). It is an interesting feature of the stochastic dynamics that one can eventually expect to get the usual quantum statistics even if the actual epistemic distribution of determinate values begins far from quantum equilibrium. [10]

---

[9] Bohm (1952), Bohm and Hiley (1993), and Bell (1984) described how to make various field quantities determinate. Vink (1993) described how to work the trick for virtually any physical property.

[10] Unlike Bohm's deterministic theory, Vink's stochastic theory does not need the distribution postulate. Because of the random mixing that is automatically generated by the dynamics, it can get by with a significantly weaker assumption.

Given that we can make the value of virtually any physical quantity determinate and describe its dynamics, Vink proposed making the value of every physical quantity determinate. The problem with this is that the Kochen–Specker theorem tells us that we can only keep the empirical predictions of the standard theory (where it makes unambiguous predictions) and make the value of every physical quantity determinate if we sacrifice the functional relationships between physical quantities.[11] The value of a particle's *position-times-momentum*, for example, would generally not be the value of its *position* multiplied by the value of its *momentum*. Indeed, the situation is worse: on Vink's theory the value of a particle's *position-squared* would generally fail to be the square of the value of its *position* squared.

In order to account for the experience of observers, however, we do not need to make every physical property determinate—we just need to make the physical property that determines our mental states determinate. Suppose that there is some physical property Q that fully determines the mental state of all observers. We do not know what Q is, but it presumably depends on such things as the details of human physiology (and the physiology of whatever other sentient beings there are), and, chances are, it is an extraordinarily complicated property (one might even expect that which property determines mental states changes over time, but to keep things simple, let's not worry about this here). Since we want mental states always to be determinate, we want Q always to have a determinate value. If Q is discrete, then we can take Vink's dynamics to describe its time-evolution. And we have a new formulation of quantum mechanics—call it the Q-theory. By hypothesis, the Q-theory provides an immediate account of the determinateness of experience, but we can also expect it to be empirically adequate in so far as Vink's dynamics makes the same predictions for Q as the standard theory does. And one might similarly

---

[11]   See Kochen and Specker (1967), and more recently Mermin (1990). Both Vink (1993) and Bub (1995*b*) have worried about the consequences of the Kochen–Specker theorem for making every physical property determinate. Vink has concluded that it does not cause any serious problems since, he argues, 'during a measurement the wave function of the quantum system effectively evolves into an eigenstate of the observable being measured, and then [the usual functional relationships hold] among any set of operators that commute with the one being measured' (1993: 1811). Bub on the other hand has concluded that the Kochen–Specker theorem does indeed pose serious problems and that one thus only ought to make determinate a single privileged physical property and the maximal set of properties that can also be made determinate given the current wave function while preserving functional relationships. A feature of Bub's proposal is that, except for the one privileged physical property, what physical properties there are changes over time.

expect that an observer's memories are typically reliable on this theory since this is true in the standard theory.

One could convert the Q-theory into a many-threads theory by taking Q to be the always determinate physical quantity and the Bohm–Bell–Vink dynamics as the connection rule that describes how local states hook together to form the histories of worlds. That is, one could take Q as the physical quantity that is determinate in all worlds and take T_{mn} as the transition probabilities for a world to go from one local state to another. Or one could convert the Q-theory into a single-mind theory by simply taking Q to provide a direct representation of the mental state of an observer, and one would have an explicit mental dynamics (even if one does not know what Q is). Or one could convert the Q theory into a many-minds theory by considering a continuous set of determinate quantities Q_r, supposing that each of these evolves according to the Bohm–Bell–Vink dynamics, and taking each to provide a direct representation of the minds associated with an observer. Each of these theories would be memory-preserving. And this last theory would look very much like AL's many-minds theory.

Because it might otherwise get lost in the list of options we are considering, I should like to say a few things about the single-mind Q-theory here. On this theory one supposes that each observer has just one mind; then while one might still think of Q as a physical quantity that, if determinate, would make all mental states determinate, rather than stipulate that this physical quantity (or any other *physical* quantity) is in fact determinate, one interprets Q as a direct representation of the *mental* state of *all* sentient observers. The Bohm-like time-evolution of Q would describe the coordinated evolution of all observers' minds. While Q is a mental parameter in this theory, it is analogous to the particle configuration in Bohm's theory: just as the configuration determines the positions of each particle and coordinates their motion in Bohm's theory, Q determines the state of each mind and coordinates the changes in these states in the single-mind Q-theory.

The single-mind Q-theory has several nice features. Of course, one immediately makes every observer's experiences and beliefs determinate by stipulating that each observer always has a determinate mental state. Since Q is understood as a mental parameter rather than a physical one, there is no physically preferred quantity. Unlike other formulations of quantum mechanics that appeal to minds, we have an explicit mental dynamics here (while we do not know what Q is, we know that it evolves according to the Bohm–Bell–Vink dynamics, and we know, among other things, that this will give us the usual statistical quantum predictions for simple experiments). And just as Bohm's theory correlates positions of distant particles to select effectively a single branch of the global state as actual, a single-mind Q-theory would do the same thing for mental states of 'distant' minds, so one does not have to worry about talking to mindless hulks. Of course, the single-mind Q-theory requires one to postulate nonphysical minds with transtemporal identities and the *mental* dynamics would fail to be covariant (in very much the same way as the particle motions in Bohm's theory fails to be covariant, but here it is a *nonphysical* property that violates relativity); but all told, it may be difficult to do much better.

7.3   *Lockwood's minds*

Lockwood agrees that the best way of understanding Everett is in terms of many minds, but he disagrees concerning many of the details of Albert and Loewer's theory. One might think of the debate between Lockwood and Albert and Loewer as being about how close a many-minds theory can get to Everett's stated position while still giving us what we expect from a good physical theory. Lockwood believes that he can eliminate both the commitment to independent minds and the commitment to the existence of a fundamentally random process (the time-evolution of the minds). Albert and Loewer believe that Lockwood ends up with a theory where there is no adequate interpretation of probability and that one can consequently have no empirical reason to accept it.

If one accepts Everett's claim that each branch is equally actual after, say, an x-spin measurement of a system initially in an eigenstate of z-spin, then, according to Lockwood, one must conclude that the observer 'is *literally* in two minds' after the measurement (1996: 166). Consequently, Lockwood concludes that there really is 'a multiplicity of distinct conscious points of view' associated with each sentient being at any given time (170). He refers to these subjectively distinct points of view as *minds*. Like Albert and Loewer, Lockwood supposes that there is a continuous infinity of minds associated with each observer and that these minds are typically in different mental states and that there is a natural measure which tells us what proportion are in what state. Lockwood insists on two things: 'first, the simultaneous existence of distinct (and indeed, of *subjectively* distinct) maximal experiences, each of which exists in a continuous infinity of identical copies; and second, the existence of a naturally preferred *measure* on sets of simultaneous maximal experiences' (182).

For Lockwood, this continuous infinity of subjectively distinct minds together make up the observer's 'multimind' or just 'Mind' for short. But on Lockwood's theory, as on Albert and Loewer's, it is presumably the fact that there is a single mind that records exactly those subjective appearances that I am having right now that accounts for my determinate experience.

So what is wrong with Albert and Loewer's theory? Lockwood complains that while Albert and Loewer's theory 'plainly succeeds in reconciling the universal occurrence, within the physical universe, of unitary evolution, with the appearance of state vector reduction in accordance with the usual statistical rules', 'the appeal to dualism, in order to make sense of quantum mechanics, strikes me as a rather desperate expedient' (1996: 176). Lockwood's goal then is to 'reap the benefits of Albert and Loewer's approach, without having either to embrace dualism or to introduce any irreducible elements of randomness into the picture' (176). His strategy is to give up the transcendental identity of minds. If one gives up the transcendental identity of minds, then one also gives up the transtemporal identity of minds, and it is pointless to worry about such things as which minds have which mental states at a time or how a mind's mental state changes over time. Since it is only in the mental dynamics that Albert and Loewer appeal to a fundamentally random process, by giving up the transcendental identity of minds Lockwood believes that he can eliminate all random processes from quantum mechanics. And finally, by giving up the transcendental identity of minds, he believes that he can recapture *complete* supervenience of the mental on the physical and thus avoid Albert and Loewer's commitment to mind–body dualism (184).

In his response to Lockwood, Loewer distinguishes between two types of many-minds theories. He calls one the Continuing Minds View (CMV). On this view there is a matter of fact concerning whether a particular mind associated with a brain's quantum state at one time is the same as a particular mind associated with the brain's quantum states at a later time. Since there are no physical facts which ground such identities over time, Loewer grants that this view is 'robustly dualist' (1996: 229). Loewer recognizes that the commitment to mind–body dualism makes the CMV inherently unattractive, but he argues that it is none the less much better than the Instantaneous Minds View (IMV), which he takes to be the view championed by Lockwood. On this view minds have no transtemporal identity, so one needs no mental facts to ground the identity of minds over time, and it seems more likely that one can avoid a dualistic theory of mind. The problem, however, is that on the IMV quantum probabilities cannot, as on the CMV, be understood as dynamical chances of observer's minds evolving to occupy various mental states. Since there is no identity of minds over time on the IMV, one cannot even talk about the time-evolution of the state of a particular mind.

Lockwood believes that his theory can provide a satisfactory account of probability. It is true that one has a natural measure over the set of instantaneous minds given by the norm-squared of the coefficients on each term in the determinate-belief expansion of the wave function. But again, since one does not have a notion of identity of a mind over time, it can presumably make no sense at all on Lockwood's view to talk about the probability of an instantaneous mind evolving from one mental state to another. Can the measure on instantaneous minds in Lockwood's theory be understood as representing quantum *probabilities*?

Albert and Loewer do not believe that it can. Loewer argues that Lockwood's measure lacks properties that are common to all measures that truly represent probabilities. Loewer believes that if it is rational for an agent to assign a probability less than one to a future event, then there must be some matter of fact about that event of which the agent is ignorant (1996: 231). But since every matter of fact on Lockwood's view is determined by the wave function alone and since an observer can know the post-measurement wave function *before* performing an x-spin measurement on a system in an eigenstate of z-spin, an observer can know every matter of fact about the post-measurement state *before performing the measurement*—on Lockwood's theory there is no matter of fact about which minds will occupy which mental states, only what proportion of minds will occupy which mental states, and this fact is fully determined by the wave function alone. But since an observer can know everything there is to know about the post-measurement state before performing the measurement, the fact that the theory assigns the number $1/2$ to the outcome x-spin up can have nothing to do with the *probability* of that outcome occurring. Thus, Loewer concludes that Lockwood's measure over instantaneous minds cannot be interpreted as a probability. Of course, on the CMV, where one has the identity of minds over time, there is a critically important matter of fact about the post-measurement state that the observer cannot know from the post-measurement *quantum* state alone: he cannot know before his measurement what the mental state of *his* mind will be after his measurement, and it is this lack of knowledge that is expressed by the observer's belief that the probability of getting x-spin up is $1/2$.

Indeed, it does seem that Lockwood has a very unusual notion of probability in mind. He claims that the probability measure on the set of simultaneous maximal experiences plays a role in his theory akin to that which elapsed time plays with respect to successive maximal experiences. Just as one might say that this pain lasted twice as long as that, Lockwood wants to be able to say that this pain 'is, superpositionally speaking, twice as *extensive* as that' when the 'probability' of this pain is twice as great as the 'probability' of that pain (182). He asks his reader to picture a two-dimensional experiential manifold, where each experience has a temporal length and a superpositional breadth: 'How bad, overall, a pain of a given, constant intensity is depends on the overall *area* which it occupies in one's experiential manifold; and this will be a function both of temporal "length" and superpositional "width" ' (182). So part of Lockwood's account of rational behaviour would presumably go something like this: since one would generally want to avoid pain, one would want to act so as to minimize the superpositional width of a painful experience just as one would want to act to minimize its temporal length, and one might thus have the measure over instantaneous minds affect one's deliberations much as the usual sort of probabilities would.

It seems to me that rather than recapturing the standard quantum probabilities, Lockwood wants to introduce a entirely new notion of probability. There is, of course, nothing wrong with a new notion of probability, but I find this one puzzling. For one thing, my personal experience provides me with no reason to suppose that experience has a superpositional dimension—as far as I can tell, my experience at every time would be accurately described by *a single term* of the superposition in the determinate-belief expansion of the global state. But further, it is difficult to make any sense of *who* experiences *what* on this theory. If the different maximal sets of experience are subjectively distinct, as Lockwood says, then how could a pain be subjectively worse for *me* if it occurred in a greater measure of elements of my *Mind*? It seems to me that since I only ever experience one mental state at a time, it must be my *mind*, not my *Mind* (or *multimind*), that accounts for my experience. But Lockwood denies that my *mind* has a transcendental identity, so to talk of *its* state or how *it* evolves would be to talk nonsense. Again, if I am right, then *my* experience is the experience of a *mind* (characterized by a single maximal mental experience) and not a *Mind* (characterized by the set of all physically possible maximal mental experiences); which means that if Lockwood gives up the transcendental identity of *minds*, then he cannot account for *my* experience. Not only is it difficult to understand Lockwood's probability measure as representing familiar probabilities, then, but it is also difficult to see how he can account for the sort of determinate experience that we take ourselves to have.

Lockwood believes that his theory is empirically equivalent to Albert and Loewer's, and since their theory is empirically adequate, he believes that one must also grant that his theory is empirically adequate. But since minds on Albert and Loewer's many-minds theory typically have reliable memories of their own past mental states and since minds on Lockwood's many-minds theory do not even have transcendental identities, there is a sense in which the two theories cannot possibly make the same empirical predictions.

While Loewer is willing to grant that the two theories predict that the same experiences will *somehow be occurring at a given time*, he argues that the two accounts cannot be *evidentially equivalent*. More specifically, he argues that if IMV is true, then one could never have empirical justification for accepting it as true; that is, Loewer argues that the IMV is *empirically incoherent*. If one agrees with Loewer's earlier conclusion that Lockwood has no adequate account of probability, then the argument that there could be no empirical justification for accepting Lockwood's theory as true is straightforward: since the measure over instantaneous minds cannot be interpreted as a probability, nothing follows from Lockwood's theory concerning the likelihood of an agent having a particular mental state at a time; consequently, that one in fact has a particular mental state can provide no evidence whatsoever for or against the theory. In contrast, one can have empirical evidence for or against the CMV since it makes specific statistical predictions about the mental state of each mind and how this state will evolve over time. That is, AL's many-minds theory is empirically coherent and Lockwood's is not.

It is Lockwood's denial of the transcendental identity of minds that makes his theory empirically incoherent. But here the problem is not, as in Bell's Everett (?) theory, that one has no empirical evidence for the theory's dynamics because one cannot in fact reliably remember one's past experience. Rather, here the claim that one can reliably remember one's past experience is just complete nonsense. On Lockwood's theory there are no minds that *could* have reliable memories of *their* past experience because minds fail to have transtemporal identities. Thus, it seems, one could never have empirical evidence for Lockwood's claim that the usual linear dynamics always correctly describes the time-evolution of the quantum-mechanical state.

The transtemporal identity of minds is a precondition for observers having reliable memories of their past experience on a many-minds theory.

This means that if the transtemporal identity of minds commits one to a robust mind–body dualism, then one can only have an empirically coherent many-minds theory (one such that if it were true, then one could have empirical evidence for accepting it as true) if one commits to a robust mind–body dualism.

Lockwood's many-minds theory can in some ways be thought of as a step away from Albert and Loewer's theory in the direction of what one might call a relative-fact theory.[12]

7.4 *Relative facts*

Everett said that 'the discontinuous "jump" into an eigenstate is . . . only a relative proposition, dependent upon our decomposition of the total

[12] Just as with many-worlds theories, there are many many-minds formulations of quantum mechanics. A recent example is developed by Chalmers in his 1996 book. Chalmers argues that 'As superpositions come to affect a subject's brain state, a number of separate minds result, corresponding to the components of the superposition. Each of these perceives a separate distinct world, corresponding to the sort of world that we perceive' (1996: 347). That separate minds would result from the superposition of brain states is supposed to be a consequence of the theory of mind that Chalmers develops in the rest of the book. Since his theory of mind is developed independently of quantum-mechanical considerations, it would be very suggestive if it did in fact predict the existence of multiple minds each with a determinate mental state when an observer's brain state is in a complex superposition of states. But I do not think that it does. If I understand Chalmers's argument correctly, it goes something like this: There is a mind associated with every physical implementation of the right sort of computation. If an observer is in a superposition of brain states that each implement a computation, then the observer is a physical system that implements each of the superposed computations. Therefore, there is one mind associated with each of the observer's superposed brain states (with the corresponding mental state). One problem with this argument is that a physical system that is in a superposition of implementing different computations would presumably not determinately implement any particular one of the superposed computations. Of course, what it takes for a physical system to implement a computation depends on exactly how one understands the implementation of computations. But if we assume the standard intepretation of states (and Chalmers provides no other way to understand physical properties) and if it is the physical dispositions of a system that determine what computations the system implements, then a superposed system would not implement the same computations that it would if it were in fact in one of the superposed states *because it would have different physical dispositions from those it would have if it were in one of the superposed states*. And it seems that the physical dispositions of a system must be at least part of what determines what computations the system implements. See Chalmers (1996: 317–18) for his account of implementation. One other highly influential many-minds theorist is Matthew Donald (1990, 1995, 1996). The differences between his approach and Albert and Loewer's, say, include the fact that Donald wants to understand our experience in terms of mental processes over time rather than in terms of a sequence of mental states and that he wants to take into account the physical properties of real brains and how they interact with their environments (1996).

wave function into the superposition, and relative to a particular chosen apparatus value' (1973: 59–60). But how could an *actual event*, the subjective experience of a real observer, depend on *a choice* of how to write the total wave function and *a choice* of a particular apparatus value from among the terms in the expression of the state when both of these would clearly be arbitrary matters of convention? One solution would be simply to deny (without introducing worlds, minds, or anything else) that there is typically any absolute matter of fact concerning the experience of any observer—that is, one might understand facts about the values of physical quantities generally and facts about an observer's experiences and beliefs more specifically as inherently relational on Everett's formulation of quantum mechanics.

There is a sense, of course, in which an observer's experience is a relative fact on each of the many-worlds and many-minds theories that we have considered so far: on the former theories an observer's experience is relative to which world he inhabits, and on the latter theories an observer's experience is relative to the mind that in fact accounts for his experiences and beliefs. Each of these theories, however, in some way privileges a particular decomposition of the wave function, and it seems to me that, given the global quantum-mechanical state, Everett took facts to be relative to *both* one's choice of a particular decomposition (since this choice is supposed to be *arbitrary*) and one's choice of a particular term in the decomposition. In order to capture this double relativity of facts, one needs a more thoroughgoing relativism than what is provided by the many-worlds and many-minds theories that we have considered so far.

Hillary Putnam at one time believed that physical facts in quantum mechanics were relative to the specification of an observer—or more specifically, relative to the specification of a particular von Neumann cut. He explained that

> when we choose to measure the 'mortality condition' of [Schrödinger's] cat (*alive* or *dead*), we choose to institute a frame *relative to which* the cat *has* a determinate property of being alive or a determinate property of being dead *and the measurement finds out which*; we are, so to speak, 'realists' *about the property we measure*; but we are not committed to realism about properties *incompatible* with the ones we measure. Relative to *this* observer *these* properties are 'real' (i.e. there to be discovered); but relative to a different observer different properties would be 'real'. There is no 'absolute' point of view. (1981: 209)

According to Putnam, there are no jumps in the physical state of a system; but rather, such talk is only an 'expression of the relativity of truth to the observer' (209). But since he seems to presuppose that there are absolute matters of fact once one specifies a particular observer, it may be that not even this sort of relativism goes far enough to capture what Everett had in mind.

Butterfield (1995), perhaps closer to the way Everett explains his theory, thinks of branches as representing a new dimension of indexicality. On this view, *branch* constitutes a parameter, like time, relative to which the same physical object might have different properties. Butterfield notes that on this view the fact that many propositions true in the apparent branch (for example, 'The pointer on the measuring device reads "x-spin up" ') are false in other branches is as straightforward as the fact that 'Anna is here at noon, 4 July 1984' is not everywhere true. A consequence of this view of branches is that even though one might be tempted to use the pronoun 'my' to specify the apparent branch, it does not necessarily mean that different branches contain different people or are associated with different minds—just as one might talk about the same person or mind at different *times*, one might talk about the same person or mind at different *branches*.

The analogy between branch and time is suggestive, but an analogy is only truly helpful when we understand at least one side of it—since we do not have a very firm grasp on the nature of time, it is unclear how one can exploit the analogy to get a satisfactory relative-fact account of our determinate experiences and beliefs. But further, it is difficult to reconcile both my naive intuitions and the results of my direct introspection concerning the basic nature of my experience with what a relative-fact theory seems to demand.

If I simultaneously have many mutually incompatible experiences, then how is the fact that these experiences are at different branches supposed to explain the fact that it always seems to me that I have *a single determinate, nondisjunctive, and nonrelative experience*? According to the analogy between time and branch indexicality, the answer must be that just as one can experience different things at different times, one can experience different things at different branches. But it seems to me that there is a significant disanalogy here. While I remember different experiences at different times, I have no evidence whatsoever of in any way experiencing or having experienced more than one branch—it seems that I, the only *I* that *I* care about, only ever experience one branch. Of course, one could simply reply that while I can know by direct introspection that I have different mental states at different times, I cannot know that I have different mental states at other branches in the same way. But this is not very satisfying when there is no empirical reason at all to believe that I ever in any way experience other branches. In any case, it is clear that if one opts for this sort of approach, one would have to have a picture of experience that is fundamentally different from my (naive) picture.[13]

Simon Saunders (1995) has a view that seems to me to be similar to the one described by Butterfield. He proposes that value-definiteness in Everett's pure wave mechanics is to be understood, as temporal tense is often understood, in purely relational terms—just as events occur 'before' only relative to the occurrence of other events, facts about the determinate value of a physical quantity are 'actual' only relative to other facts; or, perhaps better, facts about determinate values are 'actual' only relative to each other. Saunders argues that 'what is involved in the Everett procedure [for determining relative states] is poorly made out in terms of the notion of a set-theoretic collection of worlds' (1995: 236). Rather, the point of Everett's talk of relative states is that facts *are* relations; in other words, a fact concerning a physical system is typically only a fact relative to a specification of the state of every system with which it has interacted.

A relative-fact theory like this has curious consequences. Where is the Eiffel Tower? Well, there is no simple matter of fact about where it is; rather, the location of the Eiffel Tower is relative to a specification of the state of all other systems that are somehow correlated with its position. Relative to Hitler being dead, etc., the Eiffel Tower is in Paris. But relative to Hitler being alive, etc., perhaps the Eiffel Tower is in Berlin. But Hitler isn't really alive, right? Just as the location of the Eiffel Tower is a relative fact, whether Hitler died in the Second World War or not is a relative fact. Perhaps Hitler is dead relative to the Eiffel Tower being in Paris but alive relative to the Eiffel Tower being in Berlin. And so on.

So how would one account for one actually *seeing* the Eiffel Tower in Paris? Since Saunders does not want to commit to a many-worlds or many-minds ontology, one cannot even say that while the Eiffel Tower might be in Berlin in another world (or from the perspective of another mind), it is in Paris in my world (or from the perspective of my mind). The best that one could do here is to say something like this: the Eiffel Tower

---

[13] Once one starts to talk about branches as describing the local states of worlds or minds, then one is thinking of different branches in a more robust sense than one does on a relative-fact theory. This more robust sense of difference is what provides the straightforward accounts of determinate experience in the many-worlds and many-minds theories (those that allow one to make sense of the transtemporal identity of observers, that is). On a many-threads theory, for example, it is because an observer inhabits a particular world that the observer has the experiences that he does, and by dint of the fact that they inhabit different worlds, there are many different observers (though they may all look very much like me).

is in Paris relative to my seeing the Eiffel Tower in Paris. And here one cannot appeal to a matter of fact about which world one inhabits or which mind describes one's experience in order to account for one's experiences and beliefs: all records, experiences, and beliefs are, like most other facts on the theory, *essentially relative*.

There are two particularly straightforward things to say at this point. First, if there is good reason to believe that there is an absolute matter of fact concerning what one currently believes about the location of the Eiffel Tower, whether Hitler is alive, or that one is currently reading a sentence that begins with the words 'First, if there is good reason to believe', then there is good reason not to like a relative-fact theory. The proponent of such a theory has the job of trying to convince us that the rewards of accepting such a theory warrant the sacrifices one must make, but this is difficult when one must sacrifice such simple convictions as that the Eiffel Tower is in Paris and that I am currently reading. What could warrant this sort of sacrifice? The simplicity of the theory? The fact that the dynamics can be written in a covariant form? Presumably my *beliefs* about the importance of such considerations should not lead me to accept a theory that says that there is no matter of fact concerning what I *believe*.

Secondly, while a relative-fact theory is faithful to what Everett said about the fundamental relativity of states and to his view of physics as the study of quantum-mechanical correlations, it violates Everett's goal of deducing the same empirical predictions as the standard collapse formulation of quantum mechanics. The standard theory tells us that there is typically a simple matter of fact concerning what measurement result an observer ends up recording after an observation. But on the relative-fact theory there is no simple matter of fact concerning what measurement result the observer ends up recording. The two theories do not just predict different records, they predict very different sorts of records, one an absolute record and the other a relative record. If two theories must predict the same measurement records in order to make the same empirical predictions, the standard collapse theory and the relative-fact theory do not make the same empirical predictions. Indeed, it is unclear what empirical predictions the relative-fact theory makes.

Saunders believes that decoherence solves the preferred-basis problem in so far as one needs a preferred basis to account for the experiences of observers (how decoherence effects can be taken to select a physically preferred observable is something I shall discuss in the next chapter). But while he believes that decoherence considerations provide a preferred basis appropriate to explain an observer's relative experience at a time, he does not believe that they provide any sort of 'actualization' of any one decoherent history at the expense of the other decoherent histories. One is consequently left with only a branching structure of decohering histories. But for Saunders this is enough: '*The basic idea of the relational approach is that this is all that is required at the level of the fundamental equations.* What is "actual", just as what is "now", is to be understood as facts about relations' (1995: 243).

But then how does one understand the usual statistical predictions of quantum mechanics? What is the significance of the *amplitude* associated to a particular relative fact by the global wave function when all relative facts represented by the wave function are supposed to be true (as relative facts)? Why do there really seem to be chance events? And what reason is there to take the relative-fact theory as empirically adequate when we seem to get simple measurement results that exhibit the usual quantum statistics rather than the relative results that are predicted with certainty on the relative-fact theory? Concerning such questions, Saunders says, 'I equate empirical adequacy with purely formal criteria of adequacy . . . Any other desiderata will ultimately amount to an *a priori* or philosophical theory of probability' (1995: 258). That is, one should allow the relative-fact theory to set its own standard of intelligibility and empirical success.

While it is probably a good idea to allow for the possibility that a new physical theory will tell us something about how to improve our empirical evaluation of physical theories, a natural objection to allowing a physical theory to set its own standard of intelligibility is that if we altogether relinquished the right to evaluate a physical theory on extra-theoretical grounds, then we would have no way to choose a theory: we would have to take every logically consistent theory as seriously as any other (indeed, we would have to take the inconsistent theories seriously as well if we understand logical consistency itself as an external criterion). While there is no consensus among scientists and philosophers about what a satisfactory physical theory should do, most would presumably agree that it should at least be consistent with our actual experience. But since the relative-fact theory denies that there is anything that corresponds to the expression 'our actual experience', one loses even this as an external standard by which one might judge the theory's empirical adequacy.[14]

---

[14] Note that here, unlike the splitting-worlds or many-minds theories, there is not even a clear notion of *my* (local) experience. On Saunders' view there are not different experiences for different *me*s rather, there is only one *me* and facts about *me* have an *essentially* relative structure.

And one is left with little that could support a claim that there are empirical reasons for accepting the relative-fact theory.

On the other hand, the analogy between branch and time is so suggestive that it would be nice if there were some way to get a satisfactory relative-fact theory along the lines suggested by Saunders. It is just that I can see no way to reconcile the many relative experiences and beliefs that are all predicted by such a theory with certainty with the random absolute experiences and beliefs that I actually (or seem to actually) take myself to have—experiences and beliefs that exhibit the usual quantum statistics.

7.5   *Correlations without correlata*

A similar view has recently been promoted by David Mermin (1997) and Carlo Rovelli (1996).[15] The basic idea behind this view is to try to understand quantum mechanics in terms of statistical correlations *without there being any determinate correlata that the statistical correlations characterize*. I shall focus on Mermin's theory, and I shall explain the sense in which I take it to be closely related to the single-mind and many-minds theories discussed earlier.

In order to take the standard theory of quantum mechanics without the collapse dynamics to be a complete physical theory, one must accept that there are typically no determinate physical measurement records at the end of a measurement. Mermin accepts this, but he also wants to explain why this is not a serious problem. The explanation is that physics, properly understood, is about correlations and only correlations. It is not about correlations between determinate physical records nor is it about correlations between any other determinate physical properties. Rather, physics is about *correlations without correlata*. According to Mermin, *'Correlations have physical reality; that which they correlate does not'* (1997: 2).

Of course, there is something about all this that may remind one of passages in Everett. Everett, after all, held that all physical laws are correlation laws (1973: 117–18) and, at least sometimes, it seems that these

---

[15] The idea that quantum mechanics is about correlations and only correlations is not new. Indeed, as Mermin points out, the idea goes back at least to Bohr's response to the original EPR paper (Mermin 1997: 25). Also, it seems to me that Mermin's position is very much like Pauli's position as described in Ch. 2.

correlation laws were not supposed to be understood in terms of correlations between any *actual* physical properties (105). Rather, for Everett, the laws of physics concern 'internal correlations' (an expression that Mermin also uses) that are expressed in the universal wave function, and these quantum-mechanical correlations do not require there to be any determinate physical properties that are in fact correlated.

While it is difficult to figure out just what the significance of relative states were for Everett, Mermin's position is clear. For Mermin, Everett's relative states, the states that are supposed to be internally correlated, 'have no physical significance'. To think otherwise 'sends one off into the cloud-cuckoo-land of many worlds' (1997: 26).

Mermin's argument that quantum mechanics describes correlations without correlata goes something like this: If one takes the usual quantum-mechanical state to provide a complete and accurate physical description (which one should), then it follows (as a straightforward consequence of quantum mechanics) that one must also take the set of *all* internal statistical correlations to provide a complete and accurate physical description.[16] But one cannot consistently take all of these correlations to be descriptive of the actual physical state if one also takes them to be correlations between the determinate values of physical quantities (because of the various no-go theorems).[17] So quantum mechanics is about the statistical correlations *and only* the correlations—there are no correlata: 'the correlations that quantum mechanics describes prevail among quantities whose individual values are not just unknown: they have no physical reality' (Mermin 1997: 6).

But how are we supposed to understand quantum statistical correlations without correlata? Isn't a statistical correlation *necessarily* a correlation between determinate correlata? And if there are typically no determinate physical records, then how are we supposed to explain our determinate experience? Mermin recognizes such problems, but he wants to set them aside. Rather than understand statistical correlations in terms of the determinate values of correlata, he takes the notion of *correlation* to be 'one of the primitive building blocks from which an understanding of quantum

---

[16] Mermin calls this Wootter's theorem (1997: 10–11): the quantum-mechanical state of a system provides a complete physical description of the system if and only if the set of *all* internal correlations between its subsystems (for all possible resolutions of the state) provides a complete physical description of the system. As Mermin puts it: '*The quantum state of a complex system is nothing more than a concise encapsulation of the correlations among its subsystems*' (11).

[17] See Mermin (1997: 19–21) for an example of what goes wrong.

mechanics is to be constructed' (1997: 6). And concerning the problem of explaining our determinate experience he says, 'I shall take the extraordinary ability of consciousness to go beyond its own correlations with certain other subsystems to a direct perception of its own underlying correlata as a deep puzzle about the nature of consciousness' (6).

If determinate physical records are a precondition for determinate mental records, then there would typically be no determinate mental records on Mermin's theory, which presumably means that there would be no determinate experiences. So if one assumes that determinate physical records are a precondition for determinate mental records, then this theory, like the relative-fact theory in the last section, makes no determinate empirical predictions. How should one judge the empirical merits of such a theory? Or is physics, properly understood, not about empirical predictions?

It seems, however, that Mermin believes that we do in fact typically get determinate results to our measurements. In order to allow for the possibility of determinate experience in the absence of determinate physical records, Mermin stipulates that while the usual quantum-mechanical state provides a complete description of *physical reality*, it does not provide a complete description of *reality* (1997: 8). But while this distinction allows for the possibility of determinate experience, since quantum mechanics is only about physical reality and since our determinate experience is not determined by physical reality, it seems that Mermin's formulation of quantum mechanics still makes no empirical predictions concerning our determinate experience. Mermin admits that, 'While I maintain that abandoning the ability of physics to speak of correlata is a small price to pay for the recognition that it can speak simultaneously and consistently about all correlations, there remains the question of how to tie this wonderful structure of relationships down to anything particular, if physics admits of nothing particular' (29). And he says that 'at this stage I am not prepared to offer an answer' (30).

It is worth noting that there is a sense in which Mermin's theory as it stands does not even predict the right statistical *correlations* since it does not tell us how to update quantum-mechanical correlations after a measurement. Or, put another way, Mermin's theory provides no explanation why the proper quantum-mechanical correlations after a measurement seem to be those associated with a particular eigenstate of the observable being measured. This sort of worry leads Mermin to speculate that 'I suspect that our unfathomable conscious perceptions will have to enter the picture, as a way of updating correlations' (1997: 30). But it seems to me that we need not only a rule for updating the statistical correlations on measurement but a theory that predicts determinate measurement results.[18]

If one wants a formulation of quantum mechanics that makes clear empirical predictions, then one might consider adding a way of representing our actual measurement results and a description of how these results depend on the evolution of the physical state. If one keeps Mermin's distinction between *physical reality* and *reality*, then one might still take the usual quantum-mechanical state to provide a complete description of the *physical* world. But if one adds to Mermin's theory something *nonphysical* that explains why we get determinate measurements results and why the collapse postulate works so well in updating statistical correlations, then it seems to me that one ends up with something very much like a single-mind or many-minds theory. And of course, one might fix the relative-fact theory in essentially the same way.

[18] Without determinate measurement results, the statistical correlations predicted by the theory would have nothing in principle to do with our experience, so it would be difficult to say why one would care whether or not they were the empirically *right* statistical correlations.

8

MANY HISTORIES

T H E R E is a sense in which simple interference effects are destroyed when a system's environment becomes correlated with its state. This phenomenon is called *decoherence*. I shall consider three approaches here. According to one, decoherence alone explains why we get determinate records. According to another, decoherence helps one to formulate a satisfactory interpretation of Everett by selecting a *globally* preferred basis that makes the right physical facts determinate in each Everett branch. And according to the third, decoherence selects a *locally* preferred basis for each observer that makes the right physical facts determinate *from the perspective of that particular observer*.[1]

There is a long tradition of arguing that although an observer will typically end up in an entangled superposition of recording mutually incompatible measurement results, decoherence effects will destroy the interference effects and thus provide the observer with an effectively determinate record. I shall argue, however, that environmental decoherence does not by itself explain our determinate records.

A more modest use for decoherence considerations is in determining an objectively preferred set of alternative histories in the context of a many-histories (or many-threads) formulation of quantum mechanics. If decoherence considerations select a single, objectively preferred set of possible histories and if observers typically have determinate measurement records in these histories and if the theory can explain why *we* should expect *our* history to exhibit the usual quantum statistics, then one would have a many-histories theory that was better than those discussed earlier since one would not have to choose a special preferred physical property

---

[1] See Zeh (1970) for an early discussion of decoherence in the context of Everett. We have already seen, however, that Everett did not think that decoherence effects were necessary to explain our determinate experience. Further, as far as I can tell, Everett never used decoherence effects to argue for a physically preferred basis; indeed, whenever he said anything about the subject, he always maintained that one's choice of basis was entirely arbitrary. But it seems to me that whether a decoherence formulation of Everett can be made to work is more important than whether such a formulation is a historically accurate reconstruction of what he wanted. Consequently, in this chapter we will sometimes find ourselves even farther from what Everett himself actually said than usual.

as always determinate. I shall argue, however, that one encounters similar problems in choosing a preferred set of histories as one does in choosing a preferred determinate property.

An even more modest use for decoherence considerations is in specifying a rule for what physical properties are determinate for each observer individually at a time given the current quantum-mechanical state. The hope here is that such considerations will select the observer's most immediately accessible records as determinate precisely when they need to be determinate in order to account for the observer's experiences and beliefs. One would then know what needs to be added to the usual quantum-mechanical description in order to describe the observer as having a particular determinate measurement record. I believe, however, that there are rather serious problems with each of the rules that have been cooked up so far. I shall briefly mention some characteristic problems at the end of the chapter.

8.1 *Interference effects and the environment*

Consider two interference experiments.

The first experiment is the two-slit interference experiment we started with in Chapter 1. A source emits one electron per second, these travel past a barrier with two slits A and B, and strike a phosphorescent screen. Suppose that the source and barrier are such that each electron ends up in a superposition of passing through slit A and passing through slit B. Suppose also that nothing in the environment becomes correlated to an electron's position until it hits the screen. In this case, each second there will be a small flash of light somewhere on the screen showing where the electron hit. If one marks each of these points, one will eventually observe an interference pattern on the screen. The pattern will be different from what one would get by randomly blocking one of the slits and thus forcing each electron to go through one slit or the other. As we saw, such interference behaviour led to the standard interpretation of states where one concludes that an electron did not go through slit A, it did not go through slit B, it did not go through both, and it did not go through neither; rather, it was in a *superposition* of going through slit A and going through slit B.

The second experiment is similar to the first except that a thin conducting loop is placed around slit A so that a current will be induced in the loop if and only if an electron passes through slit A. In this case one *will not* observe an interference pattern on the screen; rather, the pattern one gets on this experiment will be perfectly compatible with each electron either going through slit *A* or going through slit *B* (Fig. 8.1). This second experiment shows that there is a sense in which a system no longer exhibits quantum-mechanical behaviour when its environment becomes correlated with its state (when the current in the wire loop becomes correlated with the positions of the electrons). Given our observations of the electrons at the screen, one might feel more comfortable saying that each electron passed through one slit or the other in the second experiment than one would in the first—after all, one might argue, the pattern in the second experiment is precisely what one would expect to get if each electron did in fact go through one slit or the other.

This destruction of simple interference effects by environmental correlations is called *decoherence*. The basic argument that decoherence explains our determinate experience goes like this: just as the environmentally correlated superposition in the second experiment is empirically indistinguishable from a state where the electron passes through either one slit or the other, an environmentally correlated superposition of different measurement records is empirically indistinguishable from a particular measurement record. While there is something seductive about such an argument, I shall argue that it simply does not work. But we first need to

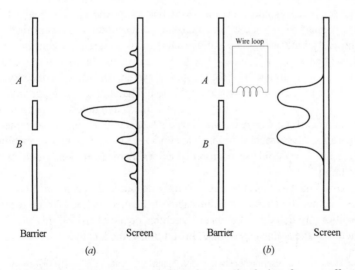

Fig. 8.1 How environmental correlations destroy simple interference effects. (*a*) The interference distribution. (*b*) With a wire loop at *A*.

consider the sense in which it would be very difficult to perform an experiment that would distinguish between a state where a measuring device recorded a superposition of results and a state where it recorded a single determinate measurement result.

8.2 The sense in which it is difficult to distinguish pure states from mixtures

Consider another Wigner's-friend story. Suppose that our friend F is ready to measure the x-spin of an electron S in an eigenstate of z-spin. Suppose further, as we have been supposing in such stories all along, that F's brain, where he records his measurement result, is perfectly isolated so that nothing in the environment of $F + S$ gets correlated with F's measurement record.[2] Since the electron is in a superposition of the x-spin states, and since F is a good observer, the usual linear dynamics predicts that F's brain will end up correlated with the x-spin of the electron: and S is initially in the state described by

$$\frac{1}{\sqrt{2}}(|\uparrow\rangle_F|\uparrow\rangle_S + |\downarrow\rangle_F|\downarrow\rangle_S). \tag{8.1}$$

Now consider an observable A of the composite system $F + S$ that has precisely this state as an eigenstate corresponding to eigenvalue $+1$ and everything orthogonal as eigenstates corresponding to eigenvalue -1. Since neither $|\uparrow\rangle_F|\uparrow\rangle_S$ nor $|\downarrow\rangle_F|\downarrow\rangle_S$ is an eigenstate of A, if one could make A-measurements of the composite system (or better, a collection of such systems), then one would be able to tell whether F's interaction with S was correctly described by the usual linear dynamics or the collapse dynamics: if a collapse occurred when F made its x-spin measurement, then an A-measurement might yield either $+1$ or -1, each with probability $1/2$; but if no collapse occurred, then an A-measurement would with certainty yield $+1$.

This suggests that one might simply go out and decide once and for all whether the friend's measurement of the electron leads to the superposition predicted by the linear dynamics or the statistical mixture predicted by the collapse postulate. In addition to the fact that F's brain is an

---

[2] This is, of course, obviously false. As Matthew Donald likes to say, real brains are warm and wet. See Donald (1996) for a rather more sophisticated model of mental states and processes than what I have been using here. Donald appeals to decoherence effects to formulate a version of quantum mechanics akin to a many-minds theory.

extraordinarily complex system, there is something else that would make it virtually impossible for us to make a measurement of $F + S$ that would distinguish between the superposition and the statistical mixture: if F is anything like a real observer, then F's environment will almost surely become correlated with F's brain state in one way or another and this will destroy the particular interference effect measured by A.

Suppose that the usual linear dynamics correctly describes the interaction between F and S but that the position of a single particle P becomes well-correlated with F's brain record of his x-spin measurement so that the final state is not

$$|\text{ideal}\rangle_{P+F+S} = |0\rangle_P \frac{1}{\sqrt{2}} (|\uparrow\rangle_F |\uparrow\rangle_S + |\downarrow\rangle_F |\downarrow\rangle_S), \qquad (8.2)$$

a state where P can be ignored in the context of measurements of $F + S$ because it is correlated with the state of neither F nor S, but

$$|\text{real}\rangle_{P+F+S} = \frac{1}{\sqrt{2}} (|0\rangle_P |\uparrow\rangle_F |\uparrow\rangle_S + |1\rangle_P |\downarrow\rangle_F |\downarrow\rangle_S), \qquad (8.3)$$

a state where $F + S$ no longer has a determinate state of its own and P can no longer be ignored when considering measurements involving F or S. The problem is that the A-measurement described above would fail to distinguish between the superposition of records represented by $|\text{real}\rangle_{P+F+S}$ and the statistical mixture of records represented by probability $1/2$ of $|0\rangle_P |\uparrow\rangle_F |\uparrow\rangle_S$ and probability $1/2$ of $|0\rangle_P |\downarrow\rangle_F |\downarrow\rangle_S$. In either case, there would be a probability of $1/2$ of getting $+1$ for the result of the A-measurement since the norm-squared of the projection of each of these three states on $|\text{ideal}\rangle_{P+F+S}$, which is the eigenstate corresponding to the eigenvalue $+1$ of the observer A for the composite system $P + F + M$, is $1/2$.

The upshot of this is that if F's brain record becomes perfectly correlated with any physical property of anything in its environment (and one would expect this to happen very quickly with any real brain record), then an A-measurement would tell one *nothing* about whether or not a collapse occurred—the stronger the correlation the less information one would get. That is, A would now be the *wrong observable* to measure to determine whether F's interaction with S was correctly described by the linear dynamics. Of course, there would be some other observable of the composite system $P + F + S$ that would in principle allow one to distinguish between $|\text{real}\rangle_{P+F+S}$ and $|0\rangle_P |\uparrow\rangle_F |\uparrow\rangle_S$ or $|1\rangle_P |\downarrow\rangle_M |\downarrow\rangle_S$ and there is always such an A-type observable (one that would distinguish between the superposition and the statistical mixture) regardless of how complicated the interaction between $F + S$ and its environment is. The problem is that one would have to know exactly what the interaction was in order to know which A-type observable to measure, but since brains are complex systems that interact with their environments in complex ways, it is extraordinarily unlikely that anyone will ever actually perform any such experiment. This is the sense in which it would be very difficult, perhaps impossible, to perform a measurement that would distinguish between a superposition of brain states and a statistical mixture of brain states.

So the argument that decoherence effects by themselves solve the measurement problem goes something like this: Since one would generally not know what systems are correlated with F's measurement record nor how they are correlated, one would generally not know which measurement to perform to distinguish between $|real\rangle_{P+F+S}$ and a state where F has recorded a determinate result. Further, any measurement that would distinguish between such states would presumably be extraordinarily difficult to perform even if one knew what it was. Consequently, it would be virtually impossible to distinguish between the superposition $|real\rangle_{P+F+S}$ and a state where F reports a single, determinate result. Thus (it is argued) for all intents and purposes $|real\rangle_{P+F+S}$ describes a state where F has recorded a single, determinate result.

But even if it were *impossible* for us to perform an experiment that would distinguish between $|real\rangle_{P+F+S}$ and a state where the observer has recorded a determinate result, it would not follow that F has recorded a single, determinate result when he is in the state $|real\rangle_{P+F+S}$, nor would it follow that there is a particular result that it somehow seems to F that he got.[3] Further, the difficulty of making appropriate A-type measurements is presumably not the sort of thing that *could* explain F's belief that he got a determinate result. An observer typically believes that he recorded a determinate result not because it is difficult to perform an A-type measurement that would distinguish between his brain being in a superposition of recording different results and being in a statistical mixture but *because he knows what result he in fact recorded.*[4]

---

[3] A proponent of the bare theory might argue that it would seem to F that he got *some* determinate result, but the bare theory predicts that F will in fact fail to have determinate measurement record, and this has nothing to do with decoherence anyway.

[4] That is, we do not believe that measurements yield determinate results because of the difficulty of A-type measurements; rather, as Albert put it, this 'is the kind of thing we learn by means of direct introspection, by merely knowing that there are matters of fact about what our beliefs are' (1992: 92 n.).

In order for the theory to predict that an observer typically records a determinate result *the theory must describe the observer as in some sense having a single determinate measurement result*. But a state like $|real\rangle_{P+F+S}$ cannot be understood as a complete description of an observer with a determinate measurement record because it does not tell us what the measurement record *is*. Moreover, if he does not have one to begin with, then further correlations between the observer and his environment will do nothing whatsoever to endow him with a determinate measurement record. One has a choice: either deny that the usual quantum-mechanical state is complete and add something to the theory to specify what result the observer in fact recorded or deny that the observer recorded a single, determinate measurement result.

While the problem with the basic decoherence argument is fairly straightforward, it remains popular. And it is sometimes proposed as the solution to the determinate-experience problem in Everett's theory.

8.3 Decoherence and determinate perceptions

W. H. Zurek (1991) argues that while Everett's formulation of quantum mechanics represents an attempt to do away with Bohr's boundary between the classical and the quantum worlds, it still provides no adequate explanation of where and how it is decided what an observer actually perceives. But such an explanation is possible, he believes, if one considers interactions between the observer and his environment. Observers are macroscopic systems, and as such are extremely difficult to isolate from their environments. Consequently, the state of an observer's environment will typically quickly become correlated with the state of the observer (and his measuring device), and one should no longer expect the observer (or his measuring device) to exhibit the indefinite quantum-mechanical behaviour that results from being in a pure coherent state. According to Zurek, then, the decoherence of the observer's state 'imposes, in effect, the required embargo on the potential outcomes by allowing the observer to maintain records of alternatives and to be aware of only one branch' (1991: 37). Zurek thus believes that decoherence explains why an observer would determinately perceive a single measurement result when he was in fact part of a composite system that was in a complicated superposition of states corresponding to having recorded mutually incompatible results.

Let's consider how this is supposed to work in more detail. Suppose the x-spin of a particle S is recorded in the state of a single detector particle M. If both particles are well-isolated from their environments, then Zurek concludes, in agreement with the standard interpretation of states (the eigenvalue–eigenstate link), that there is no determinate post-measurement record of the result; rather, the resulting pure state is just the correlated superposition:

$$|\psi\rangle = 1/\sqrt{2}(|S_\uparrow\rangle|M_\uparrow\rangle + |S_\downarrow\rangle|M_\downarrow\rangle). \qquad (8.4)$$

The statistical properties of a system in this state are represented by the pure-state density matrix for S and M (see the end of Appendix A for a brief description of this notation):

$$
\begin{aligned}
\rho^p &= |\psi\rangle\langle\psi| \\
&= |\alpha|^2|S_\uparrow\rangle\langle S_\uparrow|M_\uparrow\rangle\langle M_\uparrow| + \alpha\beta^*|S_\uparrow\rangle\langle S_\downarrow|M_\uparrow\rangle\langle M_\downarrow| \\
&\quad + \alpha^*\beta|S_\downarrow\rangle\langle S_\uparrow|M_\downarrow\rangle\langle M_\uparrow| + |\beta|^2|S_\downarrow\rangle\langle S_\downarrow|M_\downarrow\rangle\langle M_\downarrow| \qquad (8.5)
\end{aligned}
$$

But after a measurement one would like the statistical properties of S and M to be represented instead by the reduced density matrix ρ^r generated by simply cancelling the off-diagonal terms of ρ^p:

$$\rho^r = |\alpha|^2|S_\uparrow\rangle\langle S_\uparrow|M_\uparrow\rangle\langle M_\uparrow| + |\beta|^2|S_\downarrow\rangle\langle S_\downarrow|M_\downarrow\rangle\langle M_\downarrow| \qquad (8.6)$$

since this describes S and M as being in a statistical mixture of S being up and M recording *up* and S being down and M recording *down*, with probabilities $|\alpha|^2$ and $|\beta|^2$ respectively. In the standard collapse theory, one can think of the collapse of the state on measurement as generating just such a statistical mixture of eigenstates of the measured observable. On a no-collapse theory, however, if we ignore possible environmental correlations, we are stuck with the pure state. Or, as Zurek puts it, 'Unitary evolution condemns every closed system to "purity"' (1991: 39). His point, of course, is that this unfortunate consequence *only* applies to closed systems.

Consider what happens in such an experiment when the pointer on the measuring device M interacts with its environment E (as would almost certainly happen with any *pointer*—after all, measuring devices are intentionally designed so that their pointers can be easily read). Suppose that M's pointer becomes correlated to the x-spin of S as described earlier and that the environment E subsequently becomes perfectly correlated with the position of M's pointer. The state of the composite system $S + M + E$ might then be represented as

$$|\psi\rangle = 1/\sqrt{2}(|S_\uparrow\rangle|M_\uparrow\rangle|E_\uparrow\rangle + |S_\downarrow\rangle|M_\downarrow\rangle|E_\downarrow\rangle). \qquad (8.7)$$

And if this ever happens, if the environment ever becomes perfectly correlated to the position of M's pointer, then, while the state of the composite system $S + M + E$ is still pure, the density matrix that describes the state of just $S + M$, the state one gets by tracing over the environment, $\text{Tr}_E |\psi\rangle\langle\psi|$, is the reduced density matrix for $S + M$ (which is how the quantum formalism represents the physical fact that once the environment becomes correlated with M's pointer, one will not be able to observe interference effects involving just the system $S + M$). That is, if the environment becomes perfectly correlated with M's measurement record, then the density matrix that one would use to represent the state of just $S + M$ describes *this* composite system as being in a statistical mixture rather than a pure state. So (if one completely ignores the distinction between proper and improper mixtures) one might conclude that the environmental correlation puts $S + M$ in precisely the same statistical mixture as generated by the collapse postulate in the standard theory, and one might then conclude that one has solved the measurement problem and provided an account of our determinate experience (since, one might argue, the standard collapse theory accounts for our determinate experience, and we have deduced the same empirical predictions by considering environmental decoherence). But there is something seriously wrong with all this: unlike the post-measurement state in the standard collapse theory, the post-measurement state here *does not describe the observer as having recorded a particular determinate result.* That is, if one believes that the observer did in fact record a determinate result, then, since neither tells us what the result *is*, neither the pure state nor the improper statistical mixture one gets by tracing over the environment can be taken as providing a complete physical description.

Zurek argues that the quantum measurement problem is caused by pure states being *too informative*: 'if the outcomes of a measurement are to become independent, with consequences that can be explored separately, a way must be found to dispose of the excess information. This disposal can be caused by an interaction with the degrees of freedom external to the system, which we will summarily refer to as "the environment"' (1991: 39). He believes that one has solved the measurement problem if one can explain why the interference information contained in the complete pure state can be ignored when one restricts one's attention to a single physical system strongly correlated with its environment, and he believes that this is explained by showing that the state of such a system can be represented by a reduced density matrix. That is, one has solved the measurement problem if one can find some way to dispose of the information contained in the complete pure state that tells us that a typical post-measurement state cannot be thought of as a statistical mixture of independent records.

But this is puzzling. If one understands the measurement problem as one of explaining why the observer perceives a single, determinate measurement result, which is what Zurek himself set out to do (he wanted to answer the question 'Why do I, the observer, perceive only one of the outcomes?' (1991: 37)), then a pure state like (8.7) does not provide too much information; rather, it provides too little. It is certainly true that if the environment becomes strongly correlated to the state of M's pointer, it will destroy certain interference effects, but it will not add anything to the state that will describe M as having recorded a particular determinate measurement result. Or, put another way, there is nothing about the pure state of $S + M + E$ (8.7) nor the state of $S + M$ as represented by the reduced density matrix (8.6) that tells us *which* result M recorded; and if one wants to explain why an observer ends up recording the determinate result that he in fact ends up recording, then one's theory must describe the observer as determinately recording *that result*. Consequently, if one takes the usual quantum-mechanical state as complete and takes the linear dynamics to be universally true, then since nothing in the quantum-mechanical state describes the observer as having recorded a particular determinate result, decoherence considerations cannot by themselves explain why the observer records the single, determinate result that he in fact records—and if there is no determinate record of a result, then there can be no determinate experience of the result.[5]

Zurek notes that the origin of the quantum measurement problem was the clash between the predictions of the Schrödinger dynamics and our awareness of determinate outcomes. This is why he wants to explain why observers end up perceiving only one of the many possible quantum alternatives. Environmental interactions single out a preferred basis for any macroscopic recording device, a 'pointer basis', and consequently, 'An effective superselection rule has emerged—decoherence prevents superpositions of the preferred basis states from persisting' (Zurek 1991: 40).

---

[5] That decoherence destroys simple interference effects does not solve the measurement problem since it does not explain the determinateness of our measurement records. It is simply wrong to suppose that a superposition of records will 'look like' a single determinate record if the simple interference effects that one might observe between the possible record states has been destroyed. In order to observe a single determinate record, there must somewhere be a single determinate record.

He concludes: that

> We have seen how classical reality emerges from the substrate of quantum physics:
> Open quantum systems are forced into states described by localized wave packets.
> These essentially classical states obey classical equations of motion, although
> with damping and fluctuations of possibly quantum origin. What else is there to
> explain? (43)

Concerning appearances, he argues that since our brains are physical systems, awareness itself becomes susceptible to physical analysis.

> In particular, the process of decoherence is bound to affect states of the brain:
> Relevant observables of individual neurons, including chemical concentrations
> and electrical potentials, are macroscopic. They obey classical dissipative equa-
> tions of motion. Thus any quantum superposition of the states of neurons will be
> destroyed far too quickly for us to become conscious of the quantum goings-on:
> Decoherence applies to our own 'state of mind.' (44)

Zurek thus concludes that an observer's determinate-belief property is the
one selected as determinate by decoherence effects and that this explains
the observer's determinate experiences and beliefs. But again if an explanation of the observer's determinate experiences and beliefs involves
describing what those experiences and beliefs are, then the second half of
this is simply false.

So what about the first half of Zurek's conclusion? After all, he does
provide a rule for selecting a particular physical quantity as determinate
for a system at a time: whatever physical property a system's environment
is in fact correlated with is determinate *for that system*. But does this rule
always select a determinate property for each observer that makes his
most immediately accessible measurement records determinate? At first
thought, at least, this seems unlikely. The problem is not that it will not
select a property, the problem is that we do not know whether it will
select the *right* property. Since the property that the rule selects as determinate for an observer depends on the interaction between the observer
and his environment and since his environment is constantly changing,
the property selected as determinate is constantly changing as well. The
brain property that is determinate just before an observer enters an air-
port metal-detector, for example, is different from the brain property that
is determinate when he is in it. Do all such environmentally selected
brain properties make an observer's records, experiences, and beliefs
determinate? We need a good argument that they do, and we need to
be sure that it is not circular—that is, if one wants to explain why *this*

rule selects a determinate physical property that guarantees determinate records, experiences, and beliefs, then presumably one cannot simply argue that it must because we do in fact make measurements that yield determinate records, experiences, and beliefs.

Moreover, Zurek's rule for selecting what physical properties are determinate for a system can only be applied to *open* systems, systems with environments, and he wants to be able to discuss the determinate properties of *closed* systems like the universe (those systems that he earlier claimed were in fact condemned to purity). Zurek refers us to Murray Gell-Mann and James Hartle's many-histories interpretation for a discussion of how decoherence considerations allow one to talk about determinate histories of the entire universe: 'The many-histories interpretation builds on the foundation of Everett's many-worlds interpretation, but with the addition of three crucial ingredients: the notion of a set of alternative coarse-grained *histories* of a quantum system, the decoherence of the histories in a set, and their approximate determinism near the effective classical limit' (Zurek 1991: 42). Since many interpretations of Everett seem to call for some natural way of characterizing a mutually exclusive and exhaustive set of alternative physically possible histories *for the entire universe*, it is certainly worth considering how Gell-Mann and Hartle select determinate properties of *closed* systems (if this is what they in fact do). I shall also briefly discuss why it has proven rather difficult to find a satisfactory rule for selecting determinate properties for even *open* systems at the end of the chapter.

8.4   *Gell-Mann and Hartle's many-histories approach*

While the notion of measurement and the distinction between observer and observed have played an important role in traditional interpretations, Murray Gell-Mann and James Hartle (henceforth GH) argue that such interpretations are inadequate for discussing cosmology.

In a theory of the whole thing there can be no fundamental division into observer and observed. Measurements and observers cannot be fundamental notions in a theory that seeks to discuss the universe when neither existed. (GH 1990: 429–30)

Of course, finding a formulation of quantum mechanics that could be applied to cosmology was one of Everett's primary concerns.

It was Everett who in 1957 first suggested how to generalize the Copenhagen framework so as to apply quantum mechanics to cosmology. His idea was to take quantum mechanics seriously and apply it to the universe as a whole. He showed how an observer could be considered part of this system and how its activities—measuring, recording, and calculating probabilities—could be described in quantum mechanics.

Yet they believe that Everett's formulation of quantum mechanics was incomplete.

It did not adequately explain the origin of the classical domain or the meaning of 'branching' that replaced the notion of measurement. It was a theory of 'many worlds' (what we would rather call 'many histories'), but it did not sufficiently explain how these were defined or how they arose. Also Everett's discussion suggests that a probability formula is somehow not needed in quantum mechanics, even though a 'measure' is introduced that, in the end, amounts to the same thing. (430)

GH see their project, then, as 'an attempt at extension, clarification, and completion of the Everett interpretation' with the ultimate goal of finding a 'coherent formulation of quantum mechanics for science as a whole, including cosmology' (430). The first thing they do is discuss what it is that we ought to expect from scientific theories generally and from quantum mechanics in particular.

GH believe that 'All predictions in science are, most honestly and most generally, the probabilistic predictions of the *time histories* of particular events in the universe' (1990: 428). A satisfactory formulation of quantum mechanics for cosmology, then, would allow one to assign probabilities to alternative histories for the universe as a whole. They believe, however, that these probabilities may be approximate in the sense that they do not need to satisfy the standard axioms of probability theory precisely (how one ought to understand such 'approximate' probabilities is something I shall discuss later). Moreover, they do not require a satisfactory formulation of quantum mechanics to assign even an approximate probability to every possible alternative history since 'In quantum mechanics not every history can be assigned a probability' (428).

Their many-histories theory, then, provides two rules: one rule that tells us what sets of alternative histories of the universe can be assigned approximate probabilities and another rule that tells us what these probabilities are. GH describe these rules in the context of the Heisenberg picture.

The Heisenberg picture differs from the Schrödinger picture in that in the former one thinks of the quantum state as constant and the observables as evolving. On the many-histories formulation, the quantum state of the universe is represented by a density operator ρ, and the yes–no observables, those observables that represent facts that are either true or false, are represented by projection operators that evolve according to the Heisenberg dynamics. The complete Hamiltonian for the universe $\hat{H}$ determines how an operator $P(t)$ corresponding to a particular *yes–no* question evolves:

$$P(t) = e^{(iHt/\hbar)} P(0) e^{(-iHt/\hbar)}. \tag{8.8}$$

GH then add histories. A history is a particular time-sequence of facts (represented by a time-sequence of projection operators) $[P_\alpha] = (P^1_{\alpha_1}(t_1), P^2_{\alpha_2}(t_2), \dots, P^n_{\alpha_n}(t_n))$. By specifying a set of alternative facts at each time (i.e. by specifying a particular time-sequence of exhaustive sets of projection operators), one determines a set of *alternative histories*.[6] Thus each history in a particular set of alternative histories describes a specific fact as being realized at each time, and the set of alternative histories describes every possible time-sequence of determinate facts *in the context of that particular set*.

On the many-histories formulation there is no single set of physically possible histories; rather, there is a different set of alternative histories for each time-sequence of alternatives one might specify. Instead of providing a single probability measure over a single set of possible alternative histories, then, as one might have expected (and as is provided by the distribution postulate in Bohm's theory, for example), the many-histories formulation provides many different *approximate* probability measures over many alternative *sets* of alternative mutually decohering histories (how one ought to understand these alternative sets of alternative histories is something we shall worry about later). The empirical content of the theory is given by the probability measure associated with each set of alternative histories; but not all sets of alternative histories can be associated with even an approximate probability measure.

In the standard two-slit experiment where an electron is in a superposition of passing through slit A and B and where nothing in its environment gets correlated to its position, there is no probability measure associated with a set of alternative histories where the electron determinately passes through A on some histories in the set and determinately passes through B on other histories in the set. If one assumes that the electron either determinately passes through A or determinately passes through B, then the probability of it striking the screen in region R is equal to the sum of the

---

[6] Where an alternative and exhaustive set of projection operators has the property $\sum_\alpha P^k_\alpha(t) = 1$, $P^k_\alpha P^k_\beta = \delta_{\alpha\beta} P^k_\alpha$. See GH (1990: 432).

probability of the electron determinately passing through A and striking R and the probability of the electron determinately passing through B and striking R. But, because of interference effects, the probability of finding the electron in R that one calculates by supposing that the electron either passed through A or passed through B is far from the observed quantum probability; that is, if ψ_A is an eigenstate of the electron having passed through A and ψ_B is an eigenstate of the electron having passed through B, then the probability density at the screen that one gets by supposing that it passed through either A or B $|\psi_A|^2 + |\psi_B|^2$ is not equal to the observed probability, the probability that one gets by supposing that it was in a superposition of passing through A and B $|\psi_A + \psi_B|^2$. This same point might be put another way: if one tries to assign probabilities to histories that describe the electron as determinately passing through either A or B in this experiment (using GH's rule for assigning probabilities), then these probabilities will not even approximately obey the standard axioms of probability theory. Hence GH conclude that such histories cannot be assigned probabilities at all.

There are, of course, experiments where one can assign *approximate* probabilities to alternative histories. In a two-slit experiment like that described above, if the state of even a single particle in the electron's environment becomes strongly correlated to the position of the electron (as happens in the second experiment at the beginning of this chapter), then one can assign approximate probabilities to histories that describe the electron as having passed through a particular slit if one is only interested in histories involving just the electron. That is, if the electron's environment becomes correlated to the slit that it passes through, then single-particle interference effects are destroyed and one can assign approximate probabilities to alternative histories for that particle; and the stronger the environmental correlation, the better behaved the probabilities. And while the electron is not in an eigenstate of passing through A and is not in an eigenstate of passing through B, if it is in an eigenstate of the *coarser-grained observable* of passing through A *or* B, then, even without supposing that it interacts with its environment, one can assign a probability, a probability of exactly one in this experiment, to a history where the electron particle passes through either A *or* B. The point here is that while the many-histories formulation does not assign probabilities to completely fine-grained sets of histories (histories that give the precise position of every particle, for example), one can always assign probabilities to sufficiently coarse-grained sets of histories. Indeed, coarse graining will always eventually yield a set containing a single (rather uninteresting) history represented by the identity operator (to which the theory assigns probability one) (GH 1990: 432–4).

In order to give the rule for when a set of coarse-grained histories can be assigned approximate probabilities and the rule that says what these probabilities are, GH define a decoherence functional $D[\text{history}_1, \text{history}_2]$ on pairs of histories in a particular set of alternative histories:

$$D\left([P_{\alpha'}], [P_{\alpha}]\right) = Tr\left[P^n_{\alpha_{n'}}(t_n) \cdots P^1_{\alpha_{1'}}(t_1)\rho P^1_{\alpha_1}(t_1) \cdots P^n_{\alpha_n}(t_n)\right],$$
(8.9)

where the projections are time-ordered with the earliest on the inside. A set of coarse-grained alternative histories, then, is said to *decohere* when the off-diagonal elements of D are sufficiently small (I shall discuss what *sufficiently small* means later) for every pair of histories in the set. This, then, gives us one of the two rules: probabilities can be assigned to *decoherent* sets of alternative coarse-grained histories. And the other rule is easily stated: for a decoherent set of histories, the probability for each history $p([P_{\alpha}])$ is given by the *diagonal* elements of D:

$$p([P_{\alpha}]) = D([P_{\alpha}], [P_{\alpha}])$$
$$= Tr\left[P^n_{\alpha_n}(t_n) \cdots P^1_{\alpha_1}(t_1)\rho P^1_{\alpha_1}(t_1) \cdots P^n_{\alpha_n}(t_n)\right]. \quad (8.10)$$

The approximate probabilities assigned to a set of decoherent histories typically fail to satisfy the standard axioms of probability theory, but the smaller the off-diagonal elements of D, the better behaved the approximate probabilities.

GH believe that decoherent sets of alternative histories give a definite meaning to Everett's talk of branches. For a given decoherent set of alternative histories, each element in the exhaustive set of projection operators at a particular time specifies an Everett branch at that time (GH 1990: 440). Suppose the density matrix ρ representing the complete quantum state of the universe is pure: $\rho = |\psi\rangle\langle\psi|$. The state $|\psi\rangle$ can be decomposed using the projection operators that define a particular set of alternative histories:

$$|\psi\rangle = \sum_{\alpha_1 \cdots \alpha_n} P^n_{\alpha_n}(t_n) \cdots P^1_{\alpha_1}(t_1)|\psi\rangle. \quad (8.11)$$

The terms on the right-hand side of the equation are approximately orthogonal because the set of histories is decoherent. GH take these terms, one term for each possible history in the particular decoherent set, to represent Everett's branches (441). The many-histories formulation of quantum mechanics then is presented as an improved version of Everett's theory where Everett's branches are understood as alternative decohering histories. But this improved version of Everett's theory is itself rather puzzling.

8.5 *Some problems*

So how are we suppose to understand *approximate* probabilities? GH tell us that such probabilities are to be understood pragmatically, and along these lines they argue that probabilities 'need obey the rules of the probability calculus only up to some standard of accuracy sufficient for all practical purposes'. And they claim that one can achieve whatever standard of accuracy one needs in a particular situation in the many-histories formulation by considering sufficiently coarse-grained histories. But the sense in which these probabilities are approximate is curious: the approximate probability that the theory assigns to an alternative history is not approximate in the sense that it is approximately equal to the actual but unknown probability for that history; rather, it is that GH believe that the probabilities associated with alternative histories in quantum mechanics must typically fail to obey the standard axioms of probability theory. That is, GH believe that 'In quantum mechanics ... it is likely that only by this means [by violating the standard axioms of probabilities theory] can probabilities be assigned to interesting histories at all' (GH 1990: 428). In particular, the probabilities assigned by the many-histories theory violate the standard sum rule—the probabilities assigned to mutually exclusive and exhaustive alternative histories typically do not add to one.[7]

If the approximate probabilities are not to be understood as approximations to the probabilities that obey the standard axioms of probability theory, then we need some other way to understand approximate probabilities here. Further, a proponent of the many-histories theory would presumably want to explain why agents who accept the theory would not end up committed to irrational action or inconsistent beliefs.

---

[7] As I understand their position, in the two-slit experiment, for example, there simply can be no quantum probabilities associated with the histories where the particle determinately passes through *A* or determinately passes through *B* that satisfy the standard axioms of probability theory. Similarly, since decoherence is a matter of degree, the approximate probabilities associated with a set of alternative decoherent histories are not approximations to actual probabilities that satisfy the standard axioms (standard probabilities that are somehow 'out there' but unknown); rather, they must be approximate probabilities in the basic sense that they do not quite satisfy the standard axioms of probability theory. If this is right, then GH implicitly rule out formulations of quantum mechanics like Bohm's theory where there is a standard probability measure over alternative trajectories in configuration space.

GH themselves worry about the logical consistency of assigning probabilities in situations where the histories do not decohere. In the context
of the standard two-slit experiment, for example, they argue that one
cannot assign probabilities to histories where the electron determinately
passes through *A* or determinately passes through *B* because 'It would
be inconsistent to do so since the correct probability sum rules would
not be satisfied' (1990: 428).[8] It is curious, then, that they do not worry
that the many-histories theory ultimately makes predictions that violate
precisely these rules. While their position is clear enough, they believe
that large violations of the standard axioms are unacceptable but that
small violations are typically necessary in order to assign probabilities
at all. But logical consistency is usually not understood as a matter of
degree.[9]

Another problem concerns how we are supposed to understand *histories* in the many-histories theory. Is only one history from a particular
decoherent set actual? If so, then the usual quantum-mechanical state is
descriptively incomplete because it does not tell us which history this is.
Or do all histories somehow exist simultaneously? But if this is right, then
why do we only experience one history? Is it because different histories
describe events in different worlds and *we* only inhabit one world (as in
the many-histories theories described earlier)?

---

[8] See Griffiths (1984) for the first discussion of consistency conditions in the context of
this sort of formulation of quantum mechanics.

[9] Another way to put the problem is to note that if an agent assigns probabilities that fail
to satisfy the standard axioms of probability theory, then he is committed to irrational action.
More specifically, one can argue that the agent would be committed to accept a bet or series
of bets where he would be guaranteed to lose money *regardless of what happens*. Such a bet
is called a Dutch book. There are various ways of making a Dutch book against an agent who
assigns probabilities to mutually exclusive and exhaustive alternatives that do not add to one.
Suppose, for example, that an agent assigns probabilities $p(a) = 0.51$ and $p(\neg a) = 0.51$ to
the mutually exclusive and exhaustive alternatives a and $\neg a$ respectively (as might happen
in GH's many-histories theory). Suppose one then offered the agent the following deal:
Pay $100; then if a occurs, you get $99, and if $\neg a$ occurs, you get $99. The agent would
presumably calculate his expected return as $[(0.51)(99) + (0.51)(99)] - \$100 = \$0.98$,
so he would accept the offer expecting to win about $1. He would, however, be guaranteed
to lose exactly $1 *regardless of which alternative is realized*. This is not an issue of making
precise measurements; if the agent is in fact committed to the alternatives being mutually
exclusive and exhaustive and having the probabilities predicted by the theory, then he is
committed to pay $1 even if no one ever looks to see what actually happened. And while
one might argue that a real agent would never have sufficient information about the global
quantum-mechanical state to reach the conclusion that he should accept an offer like this,
it would be curious if the only thing that prevented an agent from irrational action was
incomplete information.

Concerning how we are to understand histories, Gell-Mann and Hartle explain that

> The problem with the 'local realism' that Einstein would have liked is not the locality but the *realism*. Quantum mechanics describes *alternative* decohering histories and one cannot assign 'reality' simultaneously to different alternatives because they are contradictory. Everett and others have described this situation, not incorrectly, but in a way that has confused some, by saying that the histories are all 'equally real' (meaning only that quantum mechanics prefers none over another except via probabilities) and by referring to 'many worlds' instead of 'many histories'. (GH 1990: 455)

It seems, then, that GH do not think of the alternative histories in a particular decoherent set as describing actual events in different worlds. On the other hand, it is not clear that they take precisely one history in a set of alternative histories to be actual either. But if they do not, then it is difficult to understand the significance of the probabilities assigned to the various histories in a set. Or perhaps they are critical of Einstein's realism because they believe that there is no single matter of fact about which history describes the world (which may suggest that they ultimately have something like the relative-fact theory in mind).

Since it is unclear what is meant by a history, it is also unclear how the theory is supposed to account for our determinate records, experiences, and beliefs. GH explain that 'The answer to Fermi's question to one of us of why we don't see Mars spread out in a quantum superposition of different positions in its orbit is that such a superposition would quickly decohere' (1990: 445). But how exactly is this supposed to work? Since Mars interacts strongly with its environment, although the usual linear dynamics tells us that it is most likely in a complicated superposition of being pretty much everywhere, there are decoherent sets of histories where each history in the set describes Mars as having an almost definite position right now. But how does the existence of such sets account for us seeing Mars where we do? After all, there are *also* other decoherent sets where none of the histories describe Mars as having a determinate position now.[10]

Further, even if one sorts out how to understand alternative histories *within a particular set of decohering histories*, there is another problem: the many-histories theory does not provide *just one* set of alternative decohering histories; rather, it provides many mutually incompatible *sets* of

[10] Consider the identity history, for example.

alternative decohering histories (each with its own approximate probability measure). And this makes the interpretation of histories in the theory all the more difficult.[11] Also, in so far as the many-histories theory fails to select a single objectively preferred set of decohering histories where observers typically have determinate measurement records, it presents us with something very much like the preferred-basis problem. Just as we were faced with the embarrassment of having to choose a special preferred physical quantity for the sort of many-histories theories discussed earlier, we are now apparently faced with the embarrassment of having to choose a special preferred set of decohering histories.

But given the initial state of the universe and its energy properties, perhaps there is some way of selecting a single objectively preferred set of decohering histories. Whatever their ultimate interpretation of histories, GH seem to want something very much like this:

It would be a striking and deeply important fact of the universe if, among its maximal sets of decohering histories, there were one roughly equivalent group with much higher classicities than all the others. That would be *the* quasiclassical domain, completely independent of any subjective criterion, and realized within quantum mechanics by utilizing only the initial condition of the universe and the Hamiltonian of the elementary particles. (GH 1990: 454)

This would, in effect, provide us with a single, objectively privileged set of alternative histories. If we also had a clear interpretation of these histories and if the histories were such that they typically described observers as having determinate measurement records and if we had an explanation why one should expect to record the usual quantum statistics, then we would have an interesting theory.

There are, however, a couple of problems with this strategy. For one thing, we do not have an objective notion of what should count as a quasi-classical domain. GH define a quasi-classical domain to be a maximally refined decoherent set of almost classical histories, so in order to provide

---

[11] Consider what happens. Suppose we try taking exactly one history from each alternative set of alternative histories as descriptive of *our* world (the history randomly chosen from the set with the approximate probabilities given by the many-histories theory), and hope that this set of histories is analogous to different coarse-grained descriptions of the same trajectory in phase space in classical mechanics (this is similar to a suggestion made by Bob Griffiths in conversation). The problem with this, however, is that there is no reason to expect such randomly selected histories to mesh at all. That is, in order to take each such history to be genuinely descriptive of the same world one would have to take physical facts in that world to be contingent on the level of description in a striking way: whether the Eiffel Tower is in Paris or Pittsburgh might, for example, depend on whether one considers the foundations of the structure to be a part of the tower itself.

an objective standard for what it takes to be a quasi-classical domain, one must first provide objective standards for what it takes to be a *decoherent* set of histories and what it takes for a set of histories to be *almost classical* (GH 1990: 437, 445–6). But whether a particular set of alternative histories is *decoherent* or not is a matter of degree. The smaller the off-diagonal terms of D are for histories in the set, the more decoherent the set is and the better behaved the probabilities assigned to the histories by the theory. Consequently, there is no objective matter of fact about whether a particular set is or is not decoherent; rather, it is just a matter of convention that depends on what sort of histories one is interested in discussing given one's degree of tolerance for violations of the axioms of probability theory. Indeed, GH at first take precisely this line and argue that the standard of decoherence one adopts is a matter of choice given particular pragmatic considerations: 'if a standard for the probabilities is required by their use, it can be met by coarse graining until [the decoherence conditions] are satisfied at the requisite level' (437). And in this pragmatic spirit, they never try to specify a sharp criterion for when a set of histories is decoherent. If one allows for approximate probabilities at all (in GH's sense of approximate), then I cannot see how one *could* argue for any *objective* standard for decoherence since the choice of a particular standard would presumably never amount to anything more than better- or worse-behaved probabilities. Just as significant, it is unlikely that there is any objective standard for when a set of histories is *almost* classical. And if there is no objective standard for when a set of histories is decoherent and no objective standard for when a set of histories is almost classical, then there is no objective standard for when a set of histories is quasi-classical. And if there is no objective standard for when a set of histories is quasi-classical, then we ultimately have no objective standard for selecting a single, preferred set of alternative histories.

Another problem, and I think a much more serious one, concerns how the existence of one or many quasi-classical domain(s) is supposed to account for our determinate records, experiences, and beliefs. This problem is closely connected with the problem of interpreting histories in the theory.

GH think of observers as information-gathering and -utilizing systems (IGUSes), complex adaptive systems that have evolved to exploit the relative predictability of *a particular* quasi-classical domain (1990: 425–6, 454).

The reason that such systems as IGUSes exist, functioning in such a fashion, is to be sought in their evolution within the universe. It seems likely that they evolved to make predictions because it is adaptive to do so. The reason, therefore, for their focus on decohering variables is that these are the *only* variables for which predictions can be made. (454)

And if there are many quasi-classical domains, then an IGUS would some-how 'choose' or 'exploit' just one of these.

[w]e could adopt a subjective point of view, as in some traditional discussions of quantum mechanics, and say that the IGUS 'chooses' its coarse graining of histories and, therefore, 'chooses' a particular quasiclassical domain, or sub-set of such domains for further coarse graining. It would be better, however, to say that the IGUS evolves to exploit a particular quasiclassical domain or set of such domains. Then IGUSes, including human beings, occupy no spe-cial place and play no preferred role in the laws of physics. They merely utilize probabilities presented by quantum mechanics in the context of a quasiclassical domain. (454)

But this talk of observers *choosing* and *exploiting* alternative, almost clas-sical sets of histories only serves to make the status of histories even more puzzling to me. What does it mean for an observer to choose or to exploit a particular set of histories? How is this choosing and exploiting supposed to explain our determinate measurement records? Or is it? Is my experi-ence always in fact associated with only one history? If so, then why? If not, then why does it seem that it is? One would presumably want to be able to answer such questions before claiming that we have explained our determinate measurement records, experiences, or beliefs.

8.6   *Does the environment select the right determinate quantity?*

While it is difficult to see how decoherence considerations would select a single objectively privileged set of mutually exclusive and exhaustive alternative histories for *the entire universe*, perhaps such considerations allow one to give a rule for selecting determinate physical quantities for *a particular physical system* given the global state of the universe. We want a rule that we are convinced always makes determinate precisely what needs to be determinate in order to account for our determinate measurement records, and we want this rule to work in real physical situations given the imperfections and complex environmental interactions that real observers exhibit. But finding a rule that does precisely what we want it to do is difficult.

Some rules that initially look as if they ought to work may in fact make entirely the wrong properties determinate in slightly imperfect experimental situations.[12] But perhaps more puzzling is the fact that what physical quantity a decoherence rule selects as determinate will typically depend on precisely which physical system one specifies, and this can lead to incompatible determinate properties for *nested systems*: an observer might, for example, have a determinate measurement record when his brain is the specified system but not when his whole body is the specified system.[13] How would one explain an observer's determinate experience when there is no single matter of fact about what he recorded or whether he even had a determinate record? Further, even if we settle on a canonical specification of the observing system and if we find a rule that typically makes a physical quantity determinate that is close, in an appropriate sense, to a quantity that would provide an observer with determinate measurement records (which is typically the best that a decoherence rule will be able to do), then one must also somehow argue that close is good enough to explain the determinate experiences that observers in fact have.[14] And finally, one would expect that a decoherence rule would select different properties as determinate for a given observer in different environments, so even if one had precisely the right physical property determinate at a time (a property that would make the observer's mental state determinate),

[12] For the debate concerning this problem in the context of the so-called modal theories, see Albert and Loewer (1990), Dieks (1991), Albert (1992: 191–7), Ruetsche (1995, 1998), Bacciagaluppi and Hemmo (1996a), and Vermaas (1998). One would expect many of the same issues to arise in the context of trying to find a decoherence rule for selecting which quantities are determinate for a system.

[13] See Clifton (1995) for a discussion of one popular modal rule where the determinate quantities for nested systems do not mesh. My point here is that there is a similar meshing problem for the determinate properties selected by decoherence rules since what is determinate depends on one's perspective. Suppose that I measure the x-spin of a system initially in an eigenstate of z-spin, and suppose that a particular decoherence rule chooses my recording quantity (or something that would make my record determinate) as determinate for me (say, whatever property of me becomes strongly correlated with my environment); for me and the rest of California (whatever property of me and California becomes strongly correlated with *its* environment); for me, the rest of California, and the earth (whatever property becomes correlated with its environment); but not for me, California, the earth, and the solar system *because there has not been sufficient time for anything outside the solar system to get correlated with my record*. Is there any absolute sense in which there is a determinate record? Is there a preferred physical system for explaining the determinateness of my experience? What is it and why?

[14] It seems to me that whether or not such an explanation works depends on the details of one's theory of mind: in particular, it depends on the precise details concerning the relationship between physical and mental states. See Bacciagalupi and Hemmo (1996a) for an argument that is close enough.

one would expect that that property would *not* be the one selected by the rule as determinate at other times.

There is much to say about trying to explain our determinate experience by appealing to a rule that uses the global state to select determinate quantities for individual systems, but here I would just like to suggest that it may never be obvious that a particular rule does precisely what it needs to in order to account for our determinate records, experiences, or beliefs. And we certainly have no clear argument for such a rule right now.[15]

It should be clear that decoherence does not by itself solve the measurement problem. It also seems unlikely to me that a decoherence rule would select a naturally preferred basis for the entire universe that would provide a single set of mutually exclusive and exhaustive histories or worlds. And while a decoherence rule may select a physically preferred property for an *open* macroscopic system at a time, it is not clear to me that such a rule will select the right physical property for an observer— one that would make his beliefs and memories determinate. It is certainly the case that for physical systems like brains a decoherence rule would typically select a preferred property very quickly. It is also true that, for a fixed Hamiltonian describing the interaction between the brain and its environment, one would expect that the selected property would be quite stable over time. But since the Hamiltonian that describes the interaction between our brains and the world is not fixed, it seems that we do not even have stability of the selected property, let alone a guarantee of its appropriateness. Finally, even if we did have a decoherence rule that we were convinced made a physical property that was very close to the recording brain property determinate (and very close is the best that one can expect from a decoherence rule), it is not entirely clear that close is close enough. It is, then, not yet clear, at least not to me, how decoherence effects can be used to explain our determinate records, experience, or beliefs.

---

[15] For the state of the art, see Dieks and Vermaas (1998); especially Guido Bacciagaluppi's paper describing how the Bohm–Bell–Vink dynamics might be used to describe the evolution of the actually possessed properties on one version of the modal interpretation. See Frank Arntzenius's (1998) contribution to the same volume for a detailed discussion of some of the problems faced by the modal theories. See also, Meir Hemmo's (1996) thesis for more details on how one might try to use decoherence effects in the context of a many-worlds interpretation.

9

THE DETERMINATE-EXPERIENCE PROBLEM

T H E main problem in interpreting Everett is that we do not know how he intended to deduce the same appearances in pure wave mechanics as predicted by the standard collapse theory (in so far as it makes clear empirical predictions). There are significant gaps at critical points in his exposition, and what evidence we do have concerning what he wanted is often contradictory. This explains the many mutually incompatible reconstructions of Everett that apologists and critics have devised.

Consider what a theory must do in order to make the same empirical predictions as the standard collapse theory.[1] Since the standard theory predicts that an observer will typically end up with an ordinary determinate measurement record, a theory can only make the same empirical predictions if it predicts ordinary determinate records, not disjunctive records or relative records. Also, since the standard theory predicts what an observer should expect to experience in the future, a theory can only make the same empirical predictions if it allows for some notion of transtemporal identity for observers. Finally, since the standard theory predicts that an observer's records are typically reliable, a theory can only make the same empirical predictions if it also predicts reliable records.[2] There is nothing inherently wrong with a theory not making the usual quantum predictions, but one would at least want to have a good story for why and in what sense

---

[1] Note that on some reconstructions of Everett's theory, the bare and relative-fact theories in particular, there is no explicit claim by the theory's proponents that it makes even roughly the same empirical predictions as the standard theory. Such theories describe our experience as having a fundamentally different structure from what we naively suppose it to have.

[2] If one wants to say that Bell's Everett (?) theory makes the same empirical predictions as the standard theory, then one would also have to say that a theory that predicts that my mental state is always what it is *right now* makes the same empirical predictions as the standard theory. Since this is rather silly, I am appealing to a somewhat stronger notion of empirical equivalence: two theories are empirically equivalent only if they predict the same *sequence* of experiences for an observer. And I believe that this is the way that almost everyone understands empirical equivalence.

it would *seem that it did* if the theory were true. After all, the standard collapse formulation of quantum mechanics is an incredibly successful empirical theory. Whatever its problems, it would be crazy to replace it with a theory that makes very different empirical predictions or a theory that makes no clear empirical predictions at all.

But if one takes the usual deterministic, linear quantum dynamics seriously, then one can only get the standard quantum predictions at a price. Hidden-variable and modal theories require one to choose a preferred, always determinate, physical quantity, or a rule for determining such a quantity at each time, and they require one to sacrifice the principle of state completeness by adding a new parameter, the value of the determinate physical quantity. And if one wants to be able to predict the value of the determinate quantity (and one presumably does since it is this value that will account for the experiences of observers), then such theories also require one to add an auxiliary dynamics for the evolution of the possessed values.

The splitting-worlds, many-threads, and many-histories theories require one to adopt an extravagant ontology and choose a preferred physical property for the entire universe (or a rule for selecting such a property at a time) and an auxiliary dynamics or connection rule that describes how each history or world evolves (or instead of a preferred property and an auxiliary dynamics or connection rule, one could choose the preferred set of mutually exclusive and exhaustive alternative histories directly). In exchange for the extravagant ontology (extravagant because we presumably only need a theory that describes *our* history or world in order to explain *our* experience) one might argue that such a theory preserves a sort of state completeness. In order to make predictions about what we should expect, one would also want to have a measure over the alternative histories or worlds that one could interpret as the probability of each history or world being *ours*. Albert and Loewer's many-minds theory does not require a physically preferred basis, but it does require an extravagant ontology, a robust sort of mind–body dualism, and a special rule that says how individual minds evolve. The single-mind Q-theory avoids the extravagant ontology of the many-minds theory and still solves the mindless-hulk problem. But the mental dynamics on this theory is not covariant. Further, by directly appealing to minds in order to account for our determinate experience, such theories give up on the possibility of a *purely* physical explanation of quantum phenomena. It seems that to take such theories seriously would require a fundamental change in our standards for judging scientific explanations.

Since we cannot make all the physical facts determinate in quantum mechanics that one would want to have simultaneously determinate, one might try to figure out somehow which physical facts need to be determinate in order to account for our determinate records, experience, and beliefs and then cook up a theory that makes precisely these facts determinate at precisely the right times. This is presumably what it would take to formulate a hidden-variable or modal theory that would account for our experience. Ideally one would like to get the right physical quantities determinate at the right times in order to ensure determinate experiences, but one wants to do it without the theory looking *ad hoc*. And it is still unclear how to do this.

Decoherence effects do not by themselves explain our determinate experience. It also seems unlikely that decoherence considerations will provide a way to characterize a *single* set of mutually exclusive and exhaustive alternative histories or worlds. And while such considerations may very well select a preferred physical quantity for a precisely specified observer at a time, one would like to have an argument that it selects the *right* physical quantity, a physical quantity that would explain our determinate experience, regardless of the environment or what physical system counts as the observer.

I have tried to explain the puzzles that one encounters in trying to make sense of Everett's relative-state formulation of quantum mechanics (and other no-collapse theories) as carefully and as equitably as I could. While I have presented many problems with the various no-collapse formulations of quantum mechanics, my money is on the deterministic, linear quantum dynamics making the right empirical predictions all the way up. If we are ever in a position to make the appropriate sort of interference measurement on Wigner's friend, then I have no doubt whatsoever that we would see the interference effects predicted by the standard quantum dynamics. It is my commitment to the no-collapse formulations of quantum mechanics that got me interested in Everett in the first place.[3]

---

[3] I really do not have a good argument to support this opinion. The best I can do is to suggest an inductive argument on the complexity of the systems that exhibit interference effects. Fundamental particles do, atoms do, small molecules do, medium-sized molecules do, etc. Indeed, whenever we have had the technology to check for interference effects, we have found them. Without a good reason for supposing that this pattern fails at some point, it seems reasonable enough to me to suppose that it continues.

And I might as well say which no-collapse theory I like best. I am sure that we will eventually be able to do better, but the single-mind Q-theory is my current favourite. The explicit mind–body dualism and the lack of covariance (if one associates the location of an observer's mind with the location of his body) are embarrassing. But it is the explicit mind–body dualism that provides an immediate explanation of our determinate experience. It is also what allows one to avoid having to choose a single determinate *physical* quantity or a special rule for selecting a just-right determinate quantity at each time. The lack of covariance is the price one pays for solving the mindless-hulk problem without associating a continuous infinity of minds with each physical observer. On the other hand, since it is the evolution of the observer's *mental* states, not his *physical* state, that must fail to be covariant here, perhaps some sort of reconciliation is possible with relativity. My guess is that this might require one to suppose a weaker link between physical and mental states, but this is a story for another time.

APPENDIX A

THE HILBERT-SPACE FORMALISM

W H I L E this book was written with the assumption that the reader would already know something about how quantum mechanics works, there are several good reasons for an appendix on von Neumann's Hilbert-space formalism. First, von Neumann's original presentation of the mathematical formalism is historically interesting. Secondly, his condition E may be relevant to how one understands the limiting properties of no-collapse theories (see Section 4.2). And since the Hilbert-space formalism is the standard mathematics of quantum mechanics, this appendix complements the description of the standard formulation of quantum mechanics given in Chapter 2. I shall first give a rough description of what a Hilbert space is, then I shall describe von Neumann's formal characterization.

A Hilbert space is a special sort of vector space. As a standard example of a vector space consider the set of arrows one might draw on a (very large) chalkboard. Let's say that two arrows correspond to the same *vector* if they have the same length (magnitude) and point in the same compass direction. The set of all such vectors form the elements of a vector space. One can multiply a vector v by a number α to get a new vector αv that points in the same direction as v but has a length that is longer or shorter by a factor of α. One can add vectors v and w by placing the tail of w at the tip of v then drawing a line from the tail of v to the tip of w. This line picks out a vector in the space $v + w$.

If one introduces Cartesian coordinates, then one could take the set of arrows that start at the origin and end at a point (x, y) as representing the corresponding vector. On this representation, there is one vector for each pair of coordinates (x, y), and one can define scalar multiplication and vector addition in terms of these coordinates: $\alpha(x, y) = (\alpha x, \alpha y)$ and $(x_1, y_1) + (x_2, y_2) = (x_1 + x_2, y_1 + y_2)$. Given the coordinate representation, one might also define an inner product between vectors (x_1, y_1) and (x_2, y_2) that yields the number $x_1 x_2 + y_1 y_2$. This sort of vector product can be used to give a precise definition of the length of a vector: the length of a vector v is the square root of the inner product of v with itself (one can check to see whether this gives the right answer for our vector space).

This is a real-valued vector space, but one could just as easily consider a complex-valued vector space, where the scalars and vector coordinates are complex numbers (numbers of the form $a + bi$, where a and b are ordinary real numbers and $i = \sqrt{-1}$). One could then define scalar multiplication, vector addition, and an inner product for the complex-valued representation just as one does for a real-valued representation.

Once one has a vector space, one can consider operations that rotate and stretch the vectors in the space. An operator L takes each vector v and transforms it to a new vector Lv. If L is a *linear* operator, then $L(\alpha v + \beta w) = \alpha Lv + \beta Lw$. Given the above coordinate representation, every linear operator on such vectors can be represented as a 2×2 matrix. The action of the operator on a vector is then determined by the usual rules for matrix multiplication:

$$\begin{pmatrix} a_{11} & a_{12} \\ a_{21} & a_{22} \end{pmatrix} \begin{pmatrix} v_1 \\ v_2 \end{pmatrix} = \begin{pmatrix} a_{11}v_1 + a_{12}v_2 \\ a_{21}v_1 + a_{22}v_2 \end{pmatrix}. \tag{A.1}$$

A vector space does not need to be constructed from arrows or ordered pairs of numbers. The complex-valued functions of quantum mechanics (the wave function) also form a vector space. All that matters is that the objects of the space and the operations on these objects obey the vector-space rules.

A Hilbert space is a complex-valued vector space with an inner product. In order to make the rules that define such a space mathematically precise, von Neumann stipulated that a Hilbert space $\mathcal{H}$ has the following properties:

A. $\mathcal{H}$ is a linear space. One can add elements ϕ and ψ (or $|\phi\rangle$ and $|\psi\rangle$) of $\mathcal{H}$ and get a new element $\phi + \psi$ (or $|\phi\rangle + |\psi\rangle$) of the space. One can also multiply an element ψ of $\mathcal{H}$ by a number α and get a new element $\alpha\psi$ (or $\alpha|\psi\rangle$) of the space. Vector addition and scalar multiplication have the following properties for complex numbers α and β and elements of $\mathcal{H}$ ϕ, ψ, and χ:

1. Commutative law of addition: $\phi + \psi = \psi + \phi$.
2. Associative law of addition: $(\phi + \psi) + \chi = \phi + (\psi + \chi)$.
3. Distributive law of multiplication: $(\alpha + \beta)\psi = \alpha\psi + \beta\psi$ and $\alpha(\phi + \psi) = \alpha\phi + \alpha\psi$.
4. Associative law of multiplication: $(\alpha\beta)\psi = \alpha(\beta\psi)$.
5. 0 and 1: $0\psi = 0$ and $1\psi = \psi$.

B. An inner product is defined between elements of $\mathcal{H}$. The inner product of ϕ and ψ is written as $\phi\psi$ (or $\langle\phi|\psi\rangle$). It is just a complex number, not an element of $\mathcal{H}$. The inner product has the following properties:

1. Distributive law: $(\phi + \psi)\chi = \phi\chi + \psi\chi$.
2. Associative law: $(\alpha\phi)\psi = \alpha(\phi\psi)$.
3. Hermitian symmetry: $\phi\psi = (\psi\phi)^*$, where * is the complex conjugate.
4. Definite form: $\psi\psi \geq 0$, and $\psi\psi = 0$ only if $\psi = 0$.

The inner product allows one to define length and distance. The *length* of the vector ψ is defined as $\sqrt{\psi\psi}$ (it follows from Hermitian symmetry that $\psi\psi$ is real). The *distance* between ϕ and ψ is the length of the difference of the two vectors.

A given set of elements of a Hilbert space is said to be *linearly independent* if and only if no vector in the set can be represented as a linear combination of other vectors in the set (that is, no vector ψ in the set can be represented as

$a_1\psi_1 + a_2\psi_2 + \ldots$, where $\psi_1, \psi_2, \ldots$ are other elements of the set). A set of vectors forms a *basis* for $\mathcal{H}$ if and only if all the vectors in the set are elements of $\mathcal{H}$, the vectors are linearly independent, and any element of $\mathcal{H}$ can be represented as a linear combination of the vectors in the set (that is, if ψ is an element of the space, then $\psi = a_1\psi_1 + a_2\psi_2 + \cdots$, where $a_1, a_2, \ldots$ are complex numbers and $\psi_1, \psi_2, \ldots$ are elements of the basis. The *dimension* of a space M is the maximum number of linearly independent vectors.

C. $\mathcal{H}$ can be either finite-dimensional or countably infinite-dimensional.

Von Neumann also gave two topological conditions that a Hilbert space must satisfy. The first condition, *completeness*, was meant to provide a space rich enough to allow one to take limits, and the second, *separability*, was meant to prevent the space from being so large that one loses some of the structure that allows one to understand its elements as representing quantum-mechanical states:

D. $\mathcal{H}$ is complete: every Cauchy sequence in $\mathcal{H}$, every sequence where the distance between successive elements in the sequence becomes arbitrarily small, converges to an element in $\mathcal{H}$.

E. $\mathcal{H}$ is separable: there is a countable sequence of elements in $\mathcal{H}$ that is everywhere dense in $\mathcal{H}$.

Condition D is straightforward. It just guarantees that every sequence of vectors that looks as if it ought to have a limit does in fact have a limit and that the limit is in the space.

Condition E is more subtle. Separability places a limit on the size of the space. An infinite-dimensional space is separable if and only if the dimension of the space is *countable* (that is, if and only if the space has a basis whose elements can be matched up in a one-to-one way with the positive integers). If a space is separable, then we have a unique decomposition of its elements with respect to our chosen basis, which in the context of quantum mechanics is something that allows us to make physical sense of the mathematical formalism. On the other hand, by requiring separability, von Neumann loses the ability to represent states where a continuously valued physical quantity (like position or momentum) has determinate properties as elements in the Hilbert space (since this would require there to be a continuous, and hence an *uncountable*, set of mutually orthogonal vectors in the space, which means that the space could not have a countable basis). In response to this problem von Neumann suggested using discrete quantities to represent continuous quantities to the desired degree of precision. But given this, one might say that von Neumann sacrificed some of the richness of Dirac's theory (where one can express eigenstates of position, momentum, etc.) in order to present a formulation of quantum mechanics that satisfied his requirements for mathematical rigour.

Physical observables and the dynamical laws of quantum mechanics are expressed in terms of operators on the space used to represent the physical state.

An operator L on the Hilbert space $\mathcal{H}$ takes an element ψ of $\mathcal{H}$ and maps it to a new element $L\psi$ of $\mathcal{H}$. An operator L is *linear* if and only if $L(\alpha\phi + \beta\psi) = \alpha L\phi + \beta L\psi$, where α and β are complex numbers and ϕ and ψ are elements of the Hilbert space.

Two linear operators A and A^* on $\mathcal{H}$ are said to be *adjoint* if $(A\phi)\psi = \phi(A^*\psi)$. An operator A is *Hermitian* if $A^* = A$. And it is *unitary* if $UU^* = U^*U = 1$.

An element ψ of $\mathcal{H}$ is an *eigenvector* (or an *eigenfunction*, if $\mathcal{H}$ is a function space) of an operator L with eigenvalue λ (in general, a complex number) if and only if $L\psi = \lambda\psi$. An important fact for quantum mechanics is that if the operator L is Hermitian, then its eigenvalues will be real.

Dirac's notation is particularly useful for representing operators. In Dirac's notation $|\psi\rangle\langle\phi|$ is the linear operator that maps the vector $|\xi\rangle$ to the vector $\langle\phi|\xi\rangle|\psi\rangle$. The projection of $|\psi\rangle$ onto $|\phi\rangle$ is $|\phi\rangle\langle\phi|\psi\rangle = \langle\phi|\psi\rangle|\phi\rangle$. Another useful fact written in Dirac's notation is $|\langle\phi|\psi\rangle|^2 = \mathrm{Tr}(|\psi\rangle\langle\psi|\phi\rangle\langle\phi|)$, where Tr is the trace of the operator.

APPENDIX B

A CONCRETE EXAMPLE OF AN EPR EXPERIMENT IN THE CONTEXT OF THE BARE THEORY

Suppose that the spin observable X has eigenstates $|\uparrow_x\rangle_S$ and $|\downarrow_x\rangle_S$ and that the spin observable U has eigenstates $|\uparrow_u\rangle_S = \sqrt{3}/2|\uparrow_x\rangle_S - 1/2|\downarrow_x\rangle_S$ and $|\downarrow_u\rangle_S = 1/2|\uparrow_x\rangle_S + \sqrt{3}/2|\downarrow_x\rangle_S$. The reader can check the following identities: $|\uparrow_x\rangle_S = \sqrt{3}/2|\uparrow_u\rangle_S + 1/2|\downarrow_u\rangle_S$ and $|\downarrow_x\rangle_S = -1/2|\uparrow_u\rangle_S + \sqrt{3}/2|\downarrow_u\rangle_S$. Suppose further that systems S_A and S_B are initially in the EPR state given in the text and that A and B make space-like separate measurements of their respective systems. When A measures the x-spin of S_A, the standard theory predicts that the composite system will collapse to the state $|\downarrow_x\rangle_{S_A}|\uparrow_x\rangle_{S_B}$ with probability 1/2 and collapse to the state $|\uparrow_x\rangle_{S_A}|\downarrow_x\rangle_{S_B}$ with probability 1/2. This means that $p(\uparrow_x @A) = p(\downarrow_x @A) = 1/2$. If the composite system collapses to $|\downarrow_x\rangle_{S_A}|\uparrow_x\rangle_{S_B}$, then $p(\uparrow_u @B) = 3/4$ and $p(\downarrow_u @B) = 1/4$. If the composite system collapses to $|\uparrow_x\rangle_{S_A}|\downarrow_x\rangle_{S_B}$, then $p(\uparrow_u @B) = 1/4$ and $p(\downarrow_u @B) = 3/4$. So the standard theory predicts that $p(\uparrow_x @A$ and $\uparrow_u @B) = 1/8$, $p(\uparrow_x @A$ and $\downarrow_u @B) = 3/8$, $p(\downarrow_x @A$ and $\uparrow_u @B) = 3/8$, and $p(\downarrow_x @A$ and $\downarrow_u @B) = 1/8$.

So what does the bare theory predict here? After A's x-spin measurement and B's u-spin measurement, the linear dynamics tells us that the state of the composite system will be

$$\frac{1}{2\sqrt{2}}|\uparrow_u\rangle_B|\uparrow_x\rangle_A|\uparrow_u\rangle_{S_B}|\uparrow_x\rangle_{S_A} + \frac{\sqrt{3}}{2\sqrt{2}}|\uparrow_u\rangle_B|\downarrow_x\rangle_A|\uparrow_u\rangle_{S_B}|\downarrow_x\rangle_{S_A}$$

$$- \frac{\sqrt{3}}{2\sqrt{2}}|\downarrow_u\rangle_B|\uparrow_x\rangle_A|\downarrow_u\rangle_{S_B}|\uparrow_x\rangle_{S_A}$$

$$+ \frac{1}{2\sqrt{2}}|\downarrow_u\rangle_B|\downarrow_x\rangle_A|\downarrow_u\rangle_{S_B}|\downarrow_x\rangle_{S_A}. \qquad (B.1)$$

So given the general limiting property (see the last part of section 4.2), A and B will approach an eigenstate of reporting that their measurement results were randomly distributed and statistically correlated in just the way the standard theory predicts: $p(\uparrow_x @ A \text{ and} \uparrow_u @ B) = [-1/(2\sqrt{2})]^2 = 1/8$, etc., which are exactly the joint probabilities predicted by the standard collapse theory!

REFERENCES

AHARONOV, Y., and D. Z. ALBERT (1981), 'Can we Make Sense out of the Measurement Process in Relativistic Quantum Mechanics?', *Physical Review,* D24/2: 359–70.

—— and L. VAIDMAN (1996), 'About Position Measurements which do not Show the Bohmian Particle Position', in J. T. Cushing, A. Fine, and S. Goldstein (eds.), *Bohmian Mechanics and Quantum Theory: An Appraisal* (Dordrecht: Kluwer), 141–54.

ALBERT, D. Z. (1983), 'On Quantum-Mechanical Automata', *Physics Letters,* 98A/5: 249–52.

—— (1986), 'How to Take a Picture of Another Everett World', in D. M. Greenberger (ed.), *New Techniques and Ideas in Quantum Measurement Theory*, Annals of the New York Academy of Sciences (New York: New York Academy of Sciences), 498–502.

—— (1992), *Quantum Mechanics and Experience* (Cambridge, Mass.: Harvard University Press).

—— and J. A. BARRETT (1995), 'On What it Takes to be a World', *Topoi,* 14: 35–7.

—— and B. LOEWER (1988), 'Interpreting the Many Worlds Interpretation', *Synthese,* 77: 195–213.

—— —— (1989), 'Two No-Collapse Interpretations of Quantum Theory', *Noûs,* 23: 169–86.

—— —— (1990), 'Wanted Dead or Alive: Two Attempts to Solve Schrödinger's Paradox', in A. Fine, M. Forbes, and L. Wessels (eds.), *PSA 1990,* i (East Lansing, Mich.: Philosophy of Science Association), 277–85.

—— —— (1991), 'Some Alleged Solutions to the Measurement Problem', *Synthese,* 88: 87–98.

—— —— (1993), 'Tails of Schrödinger's Cat', in R. Clifton (ed.), *Perspectives on Quantum Reality* (Dordrecht: Kluwer), 81–92.

—— and H. PUTNAM (1995), 'Further Adventures of Wigner's Friend', *Topoi,* 14: 17–22.

ARNTZENIUS, F. (1994), 'Relativistic Hidden-Variable Theories?', *Erkenntnis,* 41: 207–31.

—— (1998), 'Curiouser and Curiouser: A Personal Evaluation of Modal Interpretations', in Dieks and Vermans (1998: 337–77).

BACCIAGALUPPI, G. (1996), 'Topics in the Modal Interpretation of Quantum Mechanics', Ph.D. thesis, University of Cambridge.

—— (1998), 'Bohm–Bell Dynamics in the Modal Interpretation', in Dieks and Vermaas (1998: 177–212).

—— and M. DICKSON (1996), 'Modal Interpretations with Dynamics'.

BACCIAGALUPPI, G., and M. HEMMO (1996*a*), 'Modal Interpretations, Decoherence, and Measurements', *Studies in the History and Philosophy of Modern Physics*, 27B: 239–77.

——— ——— (1996*b*), 'State Preparation in the Modal Interpretation'.

BARRETT, J. A. (1994), 'The Suggestive Properties of Quantum Mechanics without the Collapse Postulate', *Erkenntnis*, 41: 233–52.

—— (1995*a*), 'The Single-Mind and Many-Minds Formulations of Quantum Mechanics', *Erkenntnis*, 42: 89–105.

—— (1995*b*), 'The Distribution Postulate in Bohm's Theory', *Topoi*, 14: 45–54.

—— (1996), 'Empirical Adequacy and the Availability of Reliable Records in Quantum Mechanics', *Philosophy of Science*, 63: 49–64.

—— (1997), 'On Everett's Formulation of Quantum Mechanics', *Monist*, 80/1: 70–96.

—— (1998), 'On the Nature of Experience in the Bare Theory', *Synthese*, 113/3: 347–55.

BAYM, G. (1969), *Lectures on Quantum Mechanics* (Menlo Park, Calif.: Benjamin-Cummings).

BELL, J. S. (1964), 'On the Einstein–Podolsky–Rosen Paradox', *Physics*, 1: 195–200; repr. in Bell (1987: 14–21).

—— (1966), 'On the Problem of Hidden Variables in Quantum Theory', *Reviews of Modern Physics*, 38: 447–75; repr. in Bell (197: 1–13).

—— (1971), 'Introduction to the Hidden-Variable Question', in International School of Physics: Enrico Fermi, Course IL, *Foundations of Quantum Mechanics* (New York: Academic Press), 171–81; repr. in Bell (1987: 29–39).

—— (1975), 'The Theory of Local Beables', CERN-TH. 2053, July 28; repr. in *Epistemological Letters*, March 1996; *Dialectica*, 39 (1985), 86; Bell (1987: 52–62).

—— (1976*a*), 'How to Teach Special Relativity', *Progress in Scientific Culture*, 1/2; repr. in Bell (1987: 67–80).

—— (1976*b*), 'The Measurement Theory of Everett and de Broglie's Pilot Wave', in M. Flato *et al.* (eds.), *Quantum Mechanics, Determinism, Causality, and Particles* (Dordrecht: Reidel, 11–17; repr. in Bell (1987: 93–9).

—— (1981), 'Quantum Mechanics for Cosmologists', in C. Isham, R. Penrose, and D. Sciama (eds.), *Quantum Gravity*, ii (Oxford: Clarendon Press), 611–37; repr. in Bell (1987: 117–38).

—— (1982), 'On the Impossible Pilot Wave', *Foundations of Physics*, 12: 989–99; repr. in Bell (1987: 159–68).

—— (1984), 'Beables for Quantum Field Theory', CERN-TH.4035/84, 2 Aug.; repr. in Bell (1987: 173–80).

—— (1986), 'Six Possible Worlds of Quantum Mechanics', in Sture Allén (ed.), *Proceedings of the Nobel Symposium 65: Possible Worlds in Arts and Sciences* (Stockholm: Nobel Foundation); repr. in Bell (1987: 181–95).

—— (1987), *Speakable and Unspeakable in Quantum Theory* (Cambridge: Cambridge University Press).

BERNDL, K., M. DAUMER, D. DÜRR, S. GOLDSTEIN, and N. ZANGHÍ (1995), 'A Survey of Bohmian Mechanics', *Il Nuovo Cimento*, 110B/5–6: 737–50.

BERNSTEIN, J. (1991), 'King of the Quantum', *New York Review of Books*, 38/15: 61–3.

BOHM, D. (1951), *Quantum Theory* (Englewood Cliffs, NJ: Prentice-Hall).

—— (1952), 'A Suggested Interpretation of Quantum Theory in Terms of "Hidden Variables"', pts. I and II, *Physical Review*, 85: 166–79, 180–93.

—— (1953), 'Proof that Probability Density Approaches $|\psi|^2$ in Causal Interpretation of the Quantum Theory', *Physical Review*, 89: 458–66.

—— and B. J. HILEY (1993), *The Undivided Universe: An Ontological Interpretation of Quantum Theory* (London: Routledge).

BOHR, N. (1935), 'Can the Quantum-Mechanical Description of Reality be Considered Complete?', *Physical Review*, 38: 696–702.

—— (1949), 'Discussion with Einstein on Epistemological Problems in Modern Physics', in P. A. Schilpp (1959: 201–41).

BORN, M. (1926a), 'Zur Quanten Mechanik der Stossvorgänge', *Zeitschrift für Physik*, 37: 863–7.

BORN, M. (1926b), 'Quantenmechanik der Stossvorgänge', *Zeitschrift für Physik*, 38: 803–27.

—— (ed.) (1971), *The Born–Einstein Letters* (New York: Walker).

BUB, J. (1974), *The Interpretation of Quantum Mechanics* (Dordrecht: Reidel).

—— (1992), 'Quantum Mechanics without the Projection Postulate', *Foundations of Physics*, 22: 737–54.

—— (1994), 'How to Interpret Quantum Mechanics', *Erkenntnis*, 41: 253–73.

—— (1995a), 'Interference, Noncommutativity, and Determinateness in Quantum Mechanics', *Topoi*, 14: 39–43.

—— (1995b), 'Why not Take All Observables as Beables?', in D. M. Greenberger and A. Zeilinger (1995: 761–7).

—— (1997), *Interpreting the Quantum World* (Cambridge: Cambridge University Press).

—— and R. CLIFTON (1996), 'A Uniqueness Theorem for "No Collapse" Interpretations of Quantum Mechanics', *Studies in the History and Philosophy of Modern Physics*, 27: 181–219.

—— —— and B. MONTON (1997), 'The Bare Theory has No Clothes', in G. Hellman and R. Healey (1997: 32–51).

BUTTERFIELD, J. (1995), 'Worlds, Minds, and Quanta', Paper presented to the Aristotelian Society and Mind Association Joint Session, Liverpool, July.

—— (1996), 'Whither the Minds?'

CHAITIN, GREGORY J. (1987), *Algorithmic Information Theory* (Cambridge: Cambridge University Press).

CHALMERS, D. J. (1996), *The Conscious Mind* (Oxford: Oxford University Press).

CLARK, T. D., H. PRANCE, R. J. PRANCE, and T. P. SPILLER (eds.) (1991), *Macroscopic Quantum Phenomena* (Singapore: World Scientific).

CLAUSER, J. F., and A. SHIMONY (1978), 'Bell's Theorem: Experimental Tests and Implications', *Reports on Progress in Physics*, 41: 1881–1927.

CLIFTON, R. (1995), 'Why Modal Interpretations of Quantum Mechanics Must Abandon Classical Reasoning about the Physical Properties', *International Journal of Theoretical Physics*, 34: 1302–12.

—— (1996a), 'On what Being a World Takes Away'.

—— (1996b), 'The Properties of Modal Interpretations of Quantum Mechanics', *British Journal for the Philosophy of Science*, 47: 371–98.

—— (ed.) (1996c), *Perspectives on Quantum Reality* (Dordrecht: Kluwer).

CUSHING, J. T. (1994), *Quantum Mechanics: Historical Contingency and the Copenhagan Hegemony* (Chicago: University of Chicago Press).

—— (1996), 'What Measurement Problem?', in Clifton (1996c: 167–82).

DAUMER, M., D. DÜRR, S. GOLDSTEIN, N. ZANGHÍ (1996), 'Naive Realism about Operators'.

DAVIES, PAUL (ed.) (1989), *The New Physics* (Cambridge: Cambridge University Press).

D'ESPAGNAT, B. (1971), *Foundations of Quantum Mechanics*, in International School of Physics: Enrico Fermi, Course XLIX (New York: Academic Press).

—— (1976), *Conceptual Foundations Quantum Mechanics* (Reading, Mass.: Benjamin).

—— (1995), *Veiled Reality* (Reading, Mass.: Addison-Wesley).

DEUTSCH, D. (1985a), 'Quantum Theory, the Church–Turing Principle and the Universal Quantum Computer', *Proceedings of the Royal Society*, A400: 97–117.

—— (1985b), 'Quantum Theory as a Universal Physical Theory', *International Journal of Theoretical Physics*, 24: 1–41.

DEWDNEY, C., L. HARDY, and E. J. SQUIRES (1993), 'How Late Measurements of Quantum Trajectories Can Fool a Detector', *Physics Letters*, A184: 6–11.

DEWITT, B. S. (1971), 'The Many-Universes Interpretation of Quantum Mechanics', in *Foundations of Quantum Mechanics*, International School of Physics: Enrico Fermi, Course IL (New York: Academic Press); repr. in DeWitt and Graham (1973: 167–218).

—— and N. GRAHAM (eds.) (1973), *The Many-Worlds Interpretation of Quantum Mechanics* (Princeton: Princeton University Press).

DICKSON, M., and R. CLIFTON (1998), 'Lorentz-Invariance in Modal Interpretations', in Dieks and Vermaas (1998: 9–48).

DIEKS, D. G. B. J. (1991), 'On Some Alleged Difficulties in the Interpretation of Quantum Mechanics', 86/1: 77–86.

—— and P. E. VERMAAS (eds.) (1998), *The Modal Interpretation of Quantum Mechanics* (Dordrecht: Kluwer).

DIRAC, P. A. M. (1930), *The Principles of Quantum Mechanics*, 1st edn. (Oxford: Clarendon Press).

—— (1958), *The Principles of Quantum Mechanics*, 4th edn. (Oxford: Clarendon Press).

—— (1995), *The Collected Works of P. A. M. Dirac, 1924–1948*, ed. R. H. Dalitz (Cambridge: Cambridge University Press).

DONALD, M. J. (1990), 'Quantum Theory and the Brain', *Proceedings of the Royal Society*, A427: 43–93.

—— (1995), 'A Mathematical Characterization of the Physical Structure of Observers', *Foundations of Physics*, 25/4: 529–71.

—— (1996), 'On Many-Minds Interpretations of Quantum Mechanics'.

DOWKER, F., A. KENT (1996), 'On the Consistent Histories Approach to Quantum Mechanics', *Journal of Statistical Physics*, 83/5–6: 1575–1646.

DÜRR, D., S. GOLDSTEIN, and N. ZANGHÍ (1992a), 'Quantum Equilibrium and the Origin of Absolute Uncertainty', *Journal of Statistical Physics*, 67/5–6: 843–907.

—— —— —— (1992b), 'Quantum Mechanics, Randomness, and Deterministic Reality', *Physics Letters*, A172: 6–12.

—— —— —— (1993a), 'A Global Equilibrium as the Foundation of Quantum Randomness', *Foundations of Physics*, 23/5: 721–38.

—— W. FUSSENDER, S. GOLDSTEIN, and N. ZANGHÍ (1993b), 'Comment on "Surrealistic Bohm Trajectories"', *Zeitschrift für Naturforschung*, 48A: 1261–2.

EARMAN, J. (1986), *A Primer on Determinism* (Dordrecht: Boston).

—— and J. NORTON (1996), 'Infinite Pains: The Trouble with Supertasks', in A. Morton and S. Stich (eds.) (1996), *Benacerraf and his Critics* (Cambridge: Blackwell).

EINSTEIN, A. (1926), 'Letter to M. Born: 4 December 1926', in *The Born–Einstein Letters* (New York: Walker), 90–1.

—— B. PODOLSKY, and N. ROSEN (1935), 'Can Quantum-Mechanical Description of Reality be Considered Complete?', *Physical Review*, 47: 777–80; repr. in Wheeler and Zurek (1983: 138–41).

ENGLERT, B. G., M.O. SCULLY, G. SUSSMANN, and H. WALTHER (1992),' Surrealistic Bohm Trajectories', *Zeitschrift für Naturforschung*, 47A: 1175–86.

—— —— —— —— (1993), 'Reply to Comment on "Surreal Bohm Trajectories"', *Zeitschrift für Naturforschung*, 48A: 1263–4.

EVERETT, H., III (1957a), 'On the Foundations of Quantum Mechanics', Ph.D. thesis, Princeton University.

—— (1957b), ' "Relative State" Formulation of Quantum Mechanics', *Reviews of Modern Physics*, 29: 454–62; repr. in Wheeler and Zurek (1983: 318–23).

—— (1973), 'The Theory of the Universal Wave Function', in DeWitt and Graham (1973: 3–140).

FARHI, E., J. GOLDSTONE, and S. GUTMANN (1989), 'How Probability Arises in Quantum Mechanics', *Annals of Physics*, 192/2: 368–82.

Fleming, G. (1996), 'Just how Radical is Hyperplane Dependence?', in Clifton (1996a: 11–28).

GELL-MANN, M. (1979), 'What are the Building Blocks of Matter?', in D. Huff and O. Prewitt (eds.), *The Nature of the Physical Universe* (New York: Wiley).

——— and J. B. HARTLE (1990), 'Quantum Mechanics in the Light of Quantum Cosmology', in W. H. Zurek (ed.), *Complexity, Entropy, and the Physics of Information*, Proceedings of the Santa Fe Institute Studies in the Sciences of Complexity, 8 (Redwood City, Calif.: Addison-Wesley), 425–58.

GEROCH, R. (1984), 'The Everett Interpretation', *Nôus*, 18: 617–33.

GRAHAM, N. (1973), 'The Measurement of Relative Frequency', in DeWitt and Graham (1973: 229–52).

GREENBERGER, D. M., and A. ZEILINGER (eds.) (1995), *Fundamental Problems in Quantum Theory*, Annals of the New York Academy of Sciences, no. 775 (New York: New York Academy of Sciences).

GRIFFITHS, R. (1984), 'Consistent Histories and the Interpretation of Quantum Mechanics', *Journal of Statistical Physics*, 36: 219–72.

HARTLE, J. B. (1968), 'Quantum Mechanics of Individual Systems', *American Journal of Physics*, 36/8: 704–12.

HEALEY, R. (1984), 'How Many Worlds?', *Nôus*, 18: 591–616.

——— (1989), *The Philosophy of Quantum Mechanics: An Interactive Interpretation* (Cambridge: Cambridge University Press).

HEISENBERG, W. (1925), 'Über quantentheoretische Umdeutung kinematischer und mechanischer Beziehungen', *Zeitschrift für Physik*, 33: 879–93.

——— (1927), 'Über den anschaulichen Inhalt der quantentheoretischen Kinematik und Mechanik', *Zeitschrift für Physik*, 43: 172–98.

HELLMAN, G., and R. HEALEY (eds.) (1998), *Quantum Measurement: Beyond Paradox* (Minneapolis, MN: University of Minnesota Press).

HEMMO, M. (1996), 'Quantum Mechanics without Collapse: Modal Interpretations, Histories, and Many Worlds', Ph.D. thesis, University of Cambridge.

HUGHES, R. I. G. (1989), *The Structure and Interpretation of Quantum Mechanics* (Cambridge, Mass.: Harvard University Press).

INSTITUTS SOLVAY, CONSEIL DE PHYSIQUE (1928), *Électrons et photons: Rapports et discussions du cinquième Conseil de physique tenu à Bruxelles du 24 au 29 octobre 1927 sous les auspices de l'Institut International de Physique Solvay* (Paris: Gauthier-Villars).

Jammer, M. (1974), *The Philosophy of Quantum Mechanics* (New York: Wiley).

JAUCH, J. (1968), *Foundations of Quantum Mechanics* (Reading, Mass.: Addison-Wesley).

KOCHEN, S., and E. P. SPECKER (1967), 'On the Problem of Hidden Variables in Quantum Mechanics', *Journal of Mathematics and Mechanics*, 17: 59–87.

LOCKWOOD, M. (1989), *Mind, Brain, and the Quantum* (Oxford: Blackwell).

—— (1996), 'Many Minds Interpretations of Quantum Mechanics', *British Journal for the Philosophy of Science*, 47/2: 159–88.

LOEWER, B. (1996), 'Comment on Lockwood', *British Journal for the Philosophy of Science*, 47/2: 229–32.

MAUDLIN, T. (1994), *Quantum Nonlocality and Relativity* (Oxford: Blackwell).

—— (1995), 'Three Measurement Problems', *Topoi*, 14: 7–15.

MERMIN, N. D. (1990), 'Simple Unified Form for the Major No-Hidden-Variables Theorems', *Physical Review Letters*, 65: 3373–6.

—— (1997), 'What is Quantum Mechanics Trying to Tell Us?', LANL archive preprint.

OMNÉS, R. (1992), 'Consistent Interpretations of Quantum Mechanics', *Reviews of Modern Physics*, 64/2: 339–82.

PAIS, A. (1982), *Subtle is the Lord: The Science and the Life of Albert Einstein* (New York: Oxford University Press).

—— (1988), *Inward Bound: Of Matter and Forces in the Physical World* (Oxford: Oxford University Press).

PAULI, W. (1971), 'Letter to Born, 31 March 1954', in Born (1971: 221–5).

PENROSE, R., and C. J. ISHAM (eds.) (1986), *Quantum Concepts in Space and Time* (Oxford: Oxford University Press).

PRZIBRAM, K. (ed.) (1967), *Letters on Wave Mechanics* (New York: Philosophical Library).

PUTNAM, H. (1981), 'Quantum Mechanics and the Observer', *Erkenntnis*, 16/2: 193–219.

ROVELLI, C. (1996), 'Relational Quantum Mechanics', *International Journal of Theoretical Physics*, 35: 1637–78.

RUETSCHE, L. (1995), 'Measurement Error and the Albert-Loewer Problem', *Foundations of Physics Letters*, 8/4: 327–44.

—— (1998), 'How Close is "Close Enough"?', in Dieks and Vermaas (1998: 223–40).

SAUNDERS, S. (1995), 'Time, Quantum Mechanics, and Decoherence', *Synthese*, 102/2: 235–66.

SCHILPP, P. A. (1959), *Albert Einstein: Philosopher-Scientist* (New York: Harper).

SCHRÖDINGER, E. (1926a), 'Quantisierung als Eigenwertproblem', *Annalen der Physik*, 79: 361–76.

—— (1926b), 'Quantisierung als Eigenwertproblem', *Annalen der Physik*, 81: 109–39.

—— (1935), 'The Present Situation in Quantum Mechanics', trans. J. D. Trimmer, repr. in Wheeler and Zurek (1983: 152–67).

SHIMONY, A. (1986), 'Events and Processes in the Quantum World', in Penrose and Isham (1986: 182–203).

—— (1989), 'Conceptual Foundations of Quantum Mechanics', in Davies (1989: 373–95).

STEIN, H. (1984), 'The Everett Interpretation of Quantum Mechanics: Many Worlds or None?', *Nôus*, 18: 635–52.

TIPLER, F. J. (1986), 'The Many-Worlds Interpretation of Quantum Mechanics in Quantum Cosmology', in Penrose and Isham (1986: 204–14).

VAN FRAASSEN, B. C. (1991), *Quantum Mechanics: An Empiricist View* (Oxford: Oxford University Press).

VERMAAS, P. (1998), 'The Pros and Cons of the Kochen–Dieks and the Atomic Modal Interpretation', in Dieks and Vermaas (1998: 103–48).

VINK, J. C. (1993), 'Quantum Mechanics in Terms of Discrete Beables', *Physical Review*, A48: 1808–18.

VON NEUMANN, J. (1955), *Mathematical Foundations of Quantum Mechanics*, trans. R. Beyer (Princeton: Princeton University Press); first pub. in *Mathematische Grundlagen der Quantenmechanik* (Berlin: Springer, 1932).

WEINSTEIN, S. (1996), 'Undermined', *Synthese*, 106/2: 241–51.

WHEELER, J. A. (1957), 'Assessment of Everett's "Relative State" Formulation of Quantum Theory', repr. in DeWitt and Graham (1973: 151–4).

—— and W. H. ZUREK (eds.) (1983), *Quantum Theory and Measurement* (Princeton: Princeton University Press).

WIGNER, E. P. (1961), 'Remarks on the Mind–Body Question', in I. J. Good (ed.), *The Scientist Speculates* (London: Heinemann), 284–302; repr. in Wheeler and Zurek (1983: 168–81).

ZEH, H. D. (1970), 'On the Interpretation of Measurement in Quantum Theory', *Foundations of Physics*, 1: 69–76.

—— (1991), 'Decoherence and the Transition from Quantum to Classical', *Physics Today*, 44: 36–44.

ZUREK, W. H. (1993), 'Negotiating the Tricky Border between Quantum and Classical', *Physics Today*, 46: 13–15, 81–90.

INDEX